BLUE
ON
BLUE

Other Avon Books by
Geoffrey Regan

SNAFU: GREAT AMERICAN MILITARY DISASTERS

BLUE ON BLUE

A HISTORY OF FRIENDLY FIRE

GEOFFREY REGAN

AVON BOOKS NEW YORK

BLUE ON BLUE: A HISTORY OF FRIENDLY FIRE is an original publication of Avon Books. This work has never before appeared in book form.

AVON BOOKS
A division of
The Hearst Corporation
1350 Avenue of the Americas
New York, New York 10019

Published by arrangement with the author
Library of Congress Catalog Card Number: 94-23280
ISBN: 0-380-77655-3

Library of Congress Cataloging in Publication Data:

Regan, Geoffrey.
Blue on blue : a history of friendly fire / by Geoffrey Regan.
p. cm.
Includes bibliographical references and index.
1. Military history, Modern—20th century. 2. Military history.
3. Amicide (Military science) I. Title.
D431.R44 1995 94-23280
355.4'2—dc20 CIP

First Avon Books Trade Printing: April 1995

OPM 10 9 8 7 6 5 4 3 2 1

For Frank Gregory

ACKNOWLEDGMENTS

I would like to acknowledge the considerable help I received from a number of British and American military historians in the writing of this book. The greatest debt I owe to Richard Holmes, whose advice, so willingly proffered, gave me the confidence to begin working on the subject of friendly fire, and whose suggestions put me in touch with several experts in the field. Richard has written already—albeit briefly—on the subject of amicide in his books *Firing Line* and *Nuclear Warriors,* and I was able to gain much from his helpful letters. I would also like to thank Paddy Griffith, David Chandler, and Christopher Duffy for their help and advice, as well as Professor Douglas Porch from the Naval War College in Newport, Rhode Island.

I must also acknowledge my considerable debt to the work of Charles Shrader, whose pioneering study of amicide made my own efforts possible. In addition, certain secondary works have proved most helpful to me, including Eric Bergerud's recent study of the Twenty-fifth Infantry Division in Vietnam, *Red Thunder, Tropic Lightning;* Charles Whiting's *Slaughter Over Sicily;* Denis MacShane's *Friendly Fire Whitewash;* John J. Sullivan's study of the air support for Operation Cobra, and Jonathan E. Helmreich's essay on U.S. bombings of Switzerland during World War II.

In producing a book of this kind, I have depended on the assistance of a number of libraries. I would therefore like to thank the librarian of the Staff College at the Royal Military Academy, Sandhurst, for her help in finding a number of elusive sources. Thanks also must go to the librarians of the University of Southampton, the London Library, and the Hampshire Library Service.

Most of all, I would like to thank my wife, Gillian, and my two children, Andrew and Victoria, for bearing with me during difficult times and giving me the encouragement, advice, and assistance without which the whole undertaking would have been impossible.

Contents

INTRODUCTION

A study of the subject of friendly fire in history can be a humbling experience. For someone living in the last decade of the twentieth century to be confronted with so much evidence of human folly and incompetence is enough to shake the foundations of one's optimism in a better future for mankind. Friendly fire has squandered the lives of many thousands of men, notably in the present century, yet as technology has offered us even more refined and sophisticated ways of killing our fellow men we now discover that the problem of blue-on-blue, amicide—call it what you like—is actually growing by leaps and bounds. During the Gulf War, Coalition troops killed far more of one another than the enemy did—undoubtedly a first in military history. And they did so for two reasons, neither of which is very reassuring. In the first place, the technology of war has become so advanced that it has passed beyond the capability of men to use it effectively and safely. Secondly, human beings are not improving in their capabilities as fast as the machines they create and are subject to the age-old failing of carelessness. Simple human error, exacerbated by innumerable factors, from stress and fear to anger or exhilaration, from drink-induced torpor to drug-created ecstasy, has been at the root of mistakes that have cost lives, sometimes singly but often in hundreds or thousands.

Warfare is far from being as scientific as those with a vested interest in its propagation would like us to believe. My research, limited mainly to secondary sources as it has had to be by the sheer breadth of the subject, has revealed that the activity of war and the environment in which it has taken place is more chaotic than any other endeavor in which civilized man has involved himself. The sheer carnage of an ancient clash of phalanxes or a mêlée involving heavily armored men-at-arms wielding crushing and slashing weapons like halberds and maces, must have been no more scientific than an abbatoir. However one interprets the strategy and tactics of leaders like Alexander of Macedon, Epaminondas of Thebes, Pyrrhus of Epirus, Hannibal, Scipio Africanus, or Julius Caesar, one cannot avoid the fact that in the last analysis their skills served only to bring large numbers of men into a position where they had to kill or be killed, using an array of weapons far more varied, but no less barbaric, than the blades used in a slaughterhouse. Even Hannibal, the author of Cannae—that tactical marvel that has inspired commanders right up to the present day—was only maneuvering his men into a position where they could more easily butcher their Roman opponents. It was not an intellectual exercise or a war game. There was no victory until his Africans and Gauls had plied their swords and spears to good effect. There may have been science in the mind of Hannibal as he studied his maps, but for his men there was just the fear and panic, the blood and feces, of a very unscientific process: a battle.

While praising ancient commanders for their skills, military historians have tended to scorn the abilities of medieval captains, pointing out that most of their battles were little more than mêlées, great collisions of men, too clumsy to maneuver. Although such a view does scant justice to some notable Byzantine and Arab commanders, the truth is that by their nature all ancient and medieval battles were essentially mêlées, involving hand-to-hand fighting with cutting and crushing hand weapons. Even in the sixteenth century, with the advent of cannons and handguns, cultivation of the science of war could not hide the fact that the collisions of pikemen were just as bloody and no more scientific at the sharp end than they had been in ancient Greece. Battles were still chaotic and wholly unpredictable after the combatants had crossed swords.

That this is a truism I do not for a moment deny, but that randomness and pure chance operate to the extent that they do challenges the notion of warfare as a science. Who could have predicted that the most scientific, the best-educated, the most gifted British general of the nineteenth century, Sir George Pomeroy-Colley, would fall asleep while in command of his forces on Majuba Hill in 1881 and would be shot by a Boer schoolboy just twelve years old? Probably no operation of the entire First World War was as carefully planned as the British Somme offensive of July 1, 1916. All the best military brains available had been involved, yet it was a disaster. The barbed wire facing the British army had not been properly cut by the artillery, and on that one day, the British suffered over sixty thousand casualties. Friendly fire or blue-on-blue casualties may have been accidental, yet, like traffic accidents and bunker cave-ins, they were an inherent part of war. There is no more sense in concealing the truth about blue-on-blue than there is in pretending that serving soldiers will not suffer from diseases as well as from battle wounds, and that a proportion of the diseases will be self-inflicted VD.

In this book, I have not set out merely to be contentious but to try to build on the work of Charles Shrader by taking the subject further back in time to illustrate the range of friendly fire incidents that have occurred—on the ground, in the air, and at sea—throughout history. While there have been many different reasons for individual accidents in battle, human error consistently has been present in all of them, and this is something that everyone—generals included—must learn to accept. This does not mean that one need not try to improve techniques of identification to reduce the incidence of friendly fire. Acceptance instead means coming to terms with the existence of the problem, rather than attempting to hush it up or sweep it under the carpet, to use the two ghastly expressions that seem best to sum up the recent behavior of both British and American defense departments over a Gulf War blue-on-blue.

Gulf War spokesmen were wrong to suggest, or even to allow the idea to be suggested, that war can be a precise and surgical activity. As long as service chiefs are seen to condone this attitude and to allow the general public to believe it, war will always be an acceptable alternative to diplomacy. If, however, the truth is acknowledged that war is imprecise and chaotic and that smart weapons will still be unable to tell friend from foe, or civilians from soldiers, or indeed an Iraqi child from Saddam Hussein himself, then men will be less willing to follow their leaders down the road to war.

General Norman Schwarzkopf has made it quite clear that he has heard enough on the subject of friendly fire in Vietnam, Grenada, and the Gulf to last him a lifetime. In his recent autobiography, he has been at pains to say that there is really no such thing as friendly fire: No fire is friendly. Any bullet that leaves a rifle, any shell that leaves a cannon, any rocket that leaves a plane or helicopter has been designed for one thing and one thing only: to kill. To a dead or maimed soldier, it does not matter who fired the bullet that killed or disfigured him—and it never has. Of one thing we can all be sure: No smart weapon is smart enough to differentiate between friend and foe.

Friendly fire—fratricide, amicide, blue-on-blue, or whatever—has been an integral part of military life since men first became civilized enough to kill one another with something other than their bare hands. Yet getting the military profession to accept this in public is a very different matter. Nobody ever goes out of his way to acknowledge friendly fire casualties. Only the tiniest percentage of those that occur are ever reported or officially recognized. The others are shrouded, like so much else that goes wrong in wartime, in what is euphemistically called the "fog of war." These incidents are the clearest manifestation of the fallibility of the human element in warfare. Whether the incidence of friendly fire is increasing—as the percentage figures for casualties in the Gulf War would tend to suggest—or whether their details merely are more readily available to the general public, we cannot be sure. What is certain, though, is that friendly fire or, in Charles R. Shrader's memorable neologism, amicide has provided investigative journalists in both the United States and Europe with a new stick with which to beat the military behemoths. Operation Desert Storm—with its record of 23 percent of American casualties being self-inflicted and with 77 percent of U.S. combat vehicle losses resulting from friendly fire—was a high-exposure news item throughout the world for weeks in February 1991. The death of nine British soldiers in two APCs struck by American Maverick missiles imposed strains on Anglo-

American relations at a crucial stage in the struggle against Saddam Hussein. Evidence of a cover-up on both sides of the Atlantic, as revealed by Denis MacShane in his *Friendly Fire Whitewash*, shows us that however common a feature of modern warfare friendly fire has become, it is no more acceptable by those in authority now than it was in the past. Generals are now more accountable for their actions and less shielded from the glare of publicity. They have reputations to preserve, memoirs to write, political careers to pursue, and the last thing they want to hear about is more evidence of the essentially chaotic nature of warfare. Few professionals—doctors, lawyers, accountants—would enjoy being reminded like four-star generals that their failures were their own fault and their successes attributable to someone else.

In this book, I would like to acknowledge the great assistance I have enjoyed from the researches of Charles Shrader, the pioneer in the work on amicide. His notable study has provided me with many modern examples of friendly fire in an American context and has encouraged me to delve into the comparable areas of British and European military history for others. In examining the concept of friendly fire, I have been only too aware of the diverse factors that have contributed to its operation and growth. One fundamental problem of all warfare—the difficulties imposed by the environment in which a military operation takes place—has made a substantial contribution to friendly fire. Men have taken war to most parts of the world and each locale—whether the mountains of the Dolomites in World War I, the deserts of North Africa in World War II, or the jungles of Vietnam—has presented unique problems for the soldiers on both sides. As was evident in the Gulf War, identification of ground troops from their support planes was particularly difficult in the dust and sandstorms of the desert, while in Vietnam, Burma, and New Guinea, the jungle concealed both friend and foe and sometimes made differentiation impossible. In street fighting, often a feature of the battles in Normandy in 1944 or on the Eastern Front between Germans and Russians from 1942 to 1944, enemy and friendly forces could become so closely engaged that there was no battle line, and it was almost impossible for forward troops to call up artillery support without the risk—indeed the virtual certainty—that there would be friendly casualties. Clearly, the problem of identification was a vital one, and poor visibility produced by fog, low clouds, or heavy rain was bound to increase the incidence of friendly fire. Total darkness, as in the case of night operations, has inevitably played a large part in causing friendly casualties. Although modern conflicts, including those in Vietnam, the Falklands, and the Gulf, have been fought with the technological capacity to operate at night with night sights and thermal imagers, this has not prevented numerous accidents from occurring.

Another contributory factor to the incidence of friendly fire has been the nature of a military operation and the speed with which it was conducted. In the case of Operation Desert Storm, where the rapid movement of tanks and APCs across the vast areas of the Iraqi desert resulted in the intermingling of Coalition and Iraqi forces, there were many opportunities for misidentification. Decisions by pilots and junior commanders had to be taken immediately in response to the rapidly changing situation, and in such cases mistakes were inevitable. And in an age of high technology, the killing power of modern missile weapons allows for no second thoughts. The finality of the A–10 attack on the British Warrior APCs showed that. The speed of state-of-the-art fighters, equipped with a full arsenal of guns, missiles, and bombs, has resulted in the need for split-second decisions by pilots. Any hesitation can be literally fatal. In a high-tech war, well-trained flyers will have their fingers on a button with their lives depending on their speed of reaction. As Charles Shrader has observed; "In many respects modern weapons have outstripped the ability of their human users to control them, and the employment of sophisticated camouflage, electronic deception, and other modern defensive technology further compounds the problem."

In the final analysis, friendly fire is still a human problem. If Shrader is right and technology is taking—or has already taken—weapons beyond the capacity of human beings to use them, the proportion of friendly casualties in any engagement will continue to rise. Historically, however, often it was not so much that the technology was at fault but that the raw material of all human warfare—man himself—had malfunctioned. In simple terms, soldiers, like any other workers, were capable of acts of carelessness. In the past, a misdirected spear or arrow could kill as certainly as one fired purposely. There are literally thousands of examples from the Napoleonic Wars and the American Civil War of soldiers double-charging their muskets and rifles and killing themselves or their colleagues. Others have read the wrong map coordinates and called down artillery fire on themselves or other friendly units. In World War I, the French artillery expert General Percin records numerous incidents of this sort. According to his research seventy-five thousand French casualties were caused by faulty artillery support. Whereas such mistakes may be the product of the confusion of war—the fog returns—much of it is increasingly interpreted as the result of stress. It is doubtful that an eighteenth-century drill master would have allowed himself to consider the effect of stress on an individual soldier. He could not allow himself to think in individual terms. His task was to concentrate firepower in the opening moments of a battle, and to achieve this, iron discipline was necessary. As we will see, however, once confusion set in, as it did in General Braddock's famous defeat on the Monongahela in 1755, friendly casualties were inevitable as panicky soldiers lost their discipline and fired randomly. The British redcoats of that period were so thoroughly drilled that they could hardly function effectively when thrown back on their own initiative. Their training was hard but brittle. In the following century, notably during the American Civil War, we find large numbers of civilian soldiers whose training was so inadequate that they did not know how to use their weapons and killed and injured their colleagues through sheer incompetence. One need only add the stress of combat to their inadequate training to produce a cocktail of blue-on-blues.

Charles Shrader echoes General Percin in seeing a lack of coordination as a major factor in causing friendly casualties. While it is generally the common soldier—individually or en masse—who inflicts the killing blow or the fatal bullet on his comrades, the responsibility sometimes extends to the officer whose failings have created a situation in which friendly fire can take place. Poor planning is at the root of many incidents: One has only to think of the example, related by S. L. A. Marshall, of the American attack on Pork Chop Hill in Korea in 1953, where friendly troops attacking from different sides encountered each other by surprise and opened a firefight. Sometimes plans have not been coordinated properly, either between two units of the same service or else between ground troops and their artillery and air supports. A lack of information is often the cause of unnecessary casualties.

In 1755, Lieutenant Colonel James Wolfe, soon to earn immortality on the Heights of Abraham at Quebec, told his father, "I have a very mean opinion of the infantry in general. I know their discipline to be bad, and their valor precarious. They are easily put into disorder, and hard to recover out of it; they frequently kill their officers through fear and murder one another in their confusion." For Wolfe, discipline was a quintessential part of military activities, for without it, as he says, there is only chaos. While nobody would suggest that the drill of the eighteenth-century Prussian kings is the only way to prepare men for battle, one could hardly avoid the conclusion that all successful armies have been disciplined, according to their own mores. And what I mean by disciplined is efficient and able to operate within a system, or discipline, to which all have become accustomed by training. When the training has been inadequate, the discipline is easily fractured and needs to be reinforced by a system of punishments. Historically, army discipline has been concerned not just with the training of a soldier to the peak of efficiency, but it has also been seen as the most effective way of keeping men in the battle line or the firing line when most would prefer to run away and thereby avoid danger. To enforce this in most armies, it has been necessary to make examples of shirkers by shooting them *pour encourager les autres.* Both official executions for desertion or cowardice and unofficial killings, by officers on the battlefield, have been a feature of warfare throughout history. This kind of deliberate "friendly fire"—for disciplinary purposes—differs from the random or chance occurrences that make up most of the discussion in this book. Both Charles Shrader and Richard Holmes feel that this form of enforcing discipline does not really fit into the category of "amicide", as they would define it. However, I have included a brief chapter on deliberate "friendly fire", partly based on the experiences of a British general who actually boasted about the men he had killed in enforcing discipline during the First World War. Military authorities have also reacted to such "executions," whether public and official or private and spontaneous, in the same way as for unintentional friendly casualties, namely, to try to conceal the details from the public under the heading of "died in action." And it is this aspect of friendly fire—the attempt to hush up the details and hide them from the families of the bereaved—that forms the subject of the next section: the amicide incident during the Gulf War that cost the lives of nine young British soldiers.

STORM IN THE DESERT

At 1500 hours on February 26, 1991, the tanks and armored vehicles of the British First Armored Division were speeding through the Iraqi desert as part of the allied ground forces involved in Operation Desert Storm. In bad weather and poor visibility, two British Warrior armored personnel vehicles of C Company of the Royal Regiment of Fusiliers were attacked by American A–10 tankbuster aircraft firing Maverick missiles. During the attack nine British soldiers—the oldest just twenty-one years of age—were killed. The fog of war—or in this case the rain, the high winds, and the sandstorms reported to the grieving parents by the colonel in chief of the Fusiliers, HRH the Duke of Kent, as being responsible for the tragedy—had claimed yet more victims for that most ghastly of military euphemisms: friendly fire.

Yet at 1500 hours on February 26, the bad weather of the previous night had passed. The sun shone brightly, the wind had dropped, and visibility was perfect. The fog of war had returned to metaphor. Yet not—it seems—soon enough for the authorities who swung into action, determined to cover up this tragic incident. A scenario, sketched out in the minds of politicians rather than soldiers, was concocted as a damage-limitation exercise. Friendly fire was a fact of military life, yet civilians could not be expected to understand this. The idea of men being killed by mistake, by their friends and allies, was just too much for them to swallow after all the efforts that had been made to sanitize this war, with talk of surgical strikes and smart weapons. What made matters worse for a coalition army was that the Americans had killed their allies by mistake once again. Much of the criticism of the unprofessionalism of the American armed forces made by European commentators had been based on what they regarded as the "gung ho" attitude of American soldiers and airmen, with their supposed "shoot first and identify later" tactics that had caused so many friendly fire incidents in earlier conflicts, notably Grenada, Vietnam, and Korea. Relations between America and its allies would hardly be strengthened by revelations of amicide.

For more than a century, the convention had been that soldiers who died on active service were "killed in action". During the First World War, this phrase was even used at first to denote those who died in traffic accidents, those who succumbed to weapon malfunctions, and even those shot by their own officers. It was easier that way. The parents or family of dead soldiers had a right to the comforting knowledge that their son or husband's sacrifice had been worthwhile, and that it had occurred in action with the enemy. Like the euphemism "instantly" used to describe the often long-drawn-out and horrific process of dying, "killed in action" was a powerful analgesic for grieving families. How, then, should the death of the nine young British soldiers be described to their loved ones? Could bad weather excuse the American pilots? If so, then let it be used. And how did they die, purposelessly cramped inside the steel tomb of their APC or—as was to be relayed to the parents of one of the young men by his own brigade commander—heroically struggling to rescue his fellows trapped in the blazing vehicle? At moments like this, did the truth really matter? The answer—for the parents—was a resounding yes. The truth of their son's death was all they had left. The lie about the bad weather and poor visibility indicated that a cover-up was in progress.

Many people—soldiers, politicians, and diplomats—know the truth of what happened to the British APCs on February 26, 1991, yet so far none of them has been allowed to make it public knowledge. As a result, any description of the events that took place can only use evidence limited to such an extent that no conclusive findings are possible—yet further proof, if it were needed, of the old saying that in time of war, the first casualty is the truth.

Even before the start of Operation Desert Storm, British and American commanders, General Sir Peter de la Billiere and General Norman Schwarzkopf, had been acutely aware of the danger—indeed the likelihood—of friendly fire casualties. With thousands of allied tanks and APCs in contact with Iraqi armored vehicles across hundreds of miles of desert and with close air support a vital part of allied superiority, identification was bound to be a problem. The two generals had discussed identification at great length, particularly after the matter was brought to a head by the Iraqi attack on the coastal town of Khafji, in Saudi Arabia. During the battle to regain the town, some U.S. marines in an APC were attacked and killed by American aircraft. As a result, a system of orange and green recognition panels had been introduced for allied vehicles with huge inverted *V* markers—at least six feet high—painted on the sides of tanks and on their turrets. The system was tested for aerial visibility and found to be effective. Naturally, poor weather conditions could always hinder recognition, but at the outset of the fighting both Schwarzkopf and de la Billiere were satisfied that friendly casualties could be minimized if not entirely eradicated.

In the opening stages of Operation Desert Storm, the British First Armored Division, headed by the Seventh Armored Brigade, was part of the American "left hook" designed to outflank Iraqi forces by a wide sweep north, then east before closing in on Kuwait itself. The British had been allocated targets consisting of a series of large Iraqi concentrations of armor given the names of metals by the divisional commander, Major General Rupert Smith—Bronze, Copper, Brass, Zinc, Platinum, and Tungsten. Warrior APCs, in platoons of four vehicles, accompanied the Challenger battle tanks to provide infantry backup, clearing trenches and mopping up knots of Iraqi resistance. By 1400 hours on February 26, eighty Iraqi armored vehicles had been destroyed at Objective Brass and infantry of the First Battalion of the Royal Scots and the Third Battalion of the Royal Fusiliers were brought up to clear the Iraqi trenches. This was quickly achieved, and soon more than eighty British tanks and APCs carrying the eight hundred men of the Third battalion were setting off again across the wide and very flat plain toward Objective Steel. At this stage, it is important to consider the problems of identifying Warrior APCs, which, it later transpired, were misidentified by American pilots as Iraqi T54/55 tanks. The most obvious difference, and one so fundamental that it is almost impossible to confuse the two, is the size of the turret cannon. The 30mm cannon of the Warrior is short, stubby, and in length hardly reaches the end of the vehicle's body. In contrast, the long, heavy gun of the T54/55 projects far ahead of the main body, declaring itself in outline to be the main armament of a heavy battle tank rather than an armored personnel carrier. In other respects—size, outline, profile, tracks—the two vehicles are utterly dissimilar. Above all, the Warriors were carrying brightly colored fluorescent sheets as well as large, freshly painted V identification signs. Given the high-powered binoculars available to American aircrew for identification of targets, it is amazing that such a misidentification in clear weather conditions was possible.

At the head of the Third Battalion of the Fourth Armored Brigade in its drive toward Objective Steel were the thirty-seven vehicles of C Company, with 8 Platoon, commanded by Lieutenant Brett Duxbury in the lead. On arrival, troops of the Royal Engineers dismounted and placed charges to blow up six Iraqi artillery pieces, though their gunners had already fled and there seemed to be no live Iraqi soldiers in the vicinity. The four Warrior APCs had kept up a steady fire as they approached the battery of six Iraqi guns, but it was soon evident that they need expect no resistance, and they drew up about fifty meters from the guns. While the engineers worked at placing their charges, the young Fusiliers disembarked from their APCs to relieve themselves or have a smoke, but once the charges were primed, everyone was ordered back inside the vehicles to avoid casualties from shrapnel. Although Duxbury had ordered his men to close their hatches, the driver of Warrior Callsign 22 had apparently left his half-open, and this simple oversight may have saved his life. No sooner had his men returned to their vehicles than it was ripped by an explosion, flinging him out of the hatch and saving him from the flames that now engulfed the APC. As he lay wounded, his first thought was that they had struck a mine, though onlookers wondered if they had been targeted by distant Iraqi artillery.

Inside, the stricken APC shells and grenades were exploding, and four men were already dead, with others grievously wounded. As Duxbury raced to help get the survivors out of the burning APC, he ordered a second Warrior, Callsign 23, to close in to give further help, but as it did so, it too was struck by a huge explosion, killing five of its occupants and wounding others. To other units of the Third Battalion, it seemed most likely that the Iraqis were mounting a counter-attack. But the tragic truth was that there were no enemy troops within firing distance. Death had come from the skies in the shape of two Maverick missiles fired by either one or two American A–10 tankbuster aircraft. Yet in the confusion of the battlefield, this was the last thing anyone looked for, and the witnesses were far from united in their description of what happened. Where soldiers, deep within enemy territory and only recently in contact with enemy troops, had good reason to fear mines underfoot or artillery overhead, they hardly looked for danger in the skies, where from day one of the operation they had enjoyed the assurance of total superiority. Thus, there was no unanimity in what the soldiers saw as the Warriors were hit. Some of the Fusiliers present claimed they had seen a single A–10 flying low, which gave a victory waggle of its wings after scoring two hits, others claim there were two A–10s at high altitude—American reports speak of eight thousand feet—while still other Fusiliers claim to have picked up the pilot's report of his error on their radio. Curiously, nobody speaks of seeing or hearing an aircraft before the attack, which in view of the perfect visibility and the fact the vehicles had been stationary for the previous fifteen minutes is rather surprising.

The task of the A–10 tankbusters was to provide close support for ground troops in action against enemy tanks, and the desert gave pilots almost unlimited visibility. In contrast with the complex terrain of a West European battlefield, where trees, heavy vegetation, hills, and rivers and water obstacles, as well as urban obstructions like bridges, buildings, and civilian areas, might obstruct the view of enemy tanks, the Iraqi desert provided the A–10s with a perfect killing ground, as long as they were careful to identify their targets before they fired. As we have seen, Coalition armored units had anticipated the dangers of friendly fire incidents by carrying colored panels and the large, black inverted Vs painted on all vehicles. In good weather conditions and where Coalition units were not in direct combat with active Iraqi ones, it should not have been possible for mistakes to be made. But there is the ever-present element of human failing in military affairs, and while men engage in war there are always likely to be errors. On February 26, 1991, the pilots of possibly two A–10s made serious and costly mistakes in identification. The subsequent cover-up in both Britain and the United States was curiously perverse in view of the realistic view that most military commanders take of blue-on-blue: that it is an unpleasant but unavoidable fact of military life. It may well be that politicians rather than serving officers were responsible for concealing the truth from the public.

The problem facing the American ground-support planes was that the situation on the ground was so immensely fluid. There was no fixed battle front, and the speed of the allied advance was so fast that Coalition tank and armored units were spread across an enormous area of desert. Iraqi formations had been immobilized and left far behind as the American and British tanks roared northward and eastward, outflanking the Iraqi positions in Kuwait. The A–10s involved in the blue-on-blue had apparently taken off at midday on February 26 and had initially encountered the tail end of the bad weather that the units of the British First Armored Division had reported, but soon after refueling from an air tanker, they reported that the clouds were clearing and visibility was good. Once they were fueled, they contacted the air controller in one of the AWACS Boeings, whose job was to find them a target. Unfortunately, their first assignment was obscured by blowing sand, and they checked in for another target. Over two hours had already passed when they were ordered by the Air Support Operations Center to contact a forward air controller, who happened to be a British officer. He was handling the air-ground support in the area of the British First Armored Division. At this moment, something happened that may have contributed to the eventual tragedy. As well as contacting the British air controller, the A–10 pilots also got talking to an F–16 fighter pilot who was just returning from attacking an

Iraqi target. The F–16 pilot suggested that they should fly east until they came to a north/south crossroads. If they then followed the road eastward, they would find the same Iraqi vehicles that he had attacked. The A–10s, following the instructions of the F–16 pilot, set off looking for a crossroads as he had described. But the problem was, which crossroads? The desert was crisscrossed with roads, and they could not have been certain to which of these the F–16 was referring. Moreover, how close had they really been to the F–16 when they contacted him? One of the pilots claimed to have seen an F–16 close by, but there is no certainty that this was the plane whose pilot had given them the target. Nevertheless, convinced that they had at last found a potential target, the two A–10s found a suitable crossroads, complete with smoking vehicles and apparently juicy targets. According to both pilots, their attack on these Iraqi targets was unsuccessful and so they both flew south until they came upon a target consisting of about fifty vehicles. They claimed that the British air controller had assured them that there were no friendly vehicles within ten kilometers of their area of operations. Naturally, the next step should have been to identify these targets by using the binoculars with which each pilot was equipped. According to the official American version, the pilots claimed that they identified the vehicles below as Iraqi by using these binoculars and passing over the target twice to ensure correct identification. They were adamant that there were no Coalition markings on the vehicles they attacked. There was nothing now to prevent them from attacking the vehicles below, which they were convinced were Iraqi T54/55 tanks and not the much smaller British Warrior APCs. Clearly, there was no doubt in the minds of the pilots, otherwise they would have asked the British air controller for confirmation of the grid reference of the proposed target. Once they had decided not to do this, there was nothing to prevent the tragedy that followed.

In his recent book *Storm Command,* the British commander, General Sir Peter de la Billiere, recorded his own views on the disaster. As part of a politically sensitive Coalition, he knew that unity of purpose between the United States and Britain was absolutely essential. Any breach of Anglo-American relations could have been exploited by the Iraqis to weaken the resolve of the entire Coalition, and so this particular blue-on-blue, tragic and wasteful as it was, would have to be played down. In a war situation, the priority was victory through close cooperation, and whatever differences of opinion there might be must be put aside for the duration. According to the diplomatic British commander, however, while he was able to smooth things over with Norman Schwarzkopf, matters were more difficult with the U.S. air chief, Chuck Horner, who "became deeply emotional, and could not agree that the issue needed to be left open until a formal investigation had been carried out." He apparently insisted that there was no case to answer as the fault lay entirely with the British air controller, and no blame attached to the A–10 pilots. His defense of his airmen was emotional and so wide-ranging that it could not be justified by the facts. In spite of the magnificent performance by the vast majority of U.S. airmen, there had been a number of friendly fire incidents involving loss of life, and the responsibility for most of these rested fairly and squarely on the shoulders of some of Horner's pilots. So eager was Horner to defend his corner that he even resorted to blaming mines for the incident. De la Billiere records how taken aback he was by Horner's aggression. He writes: "As I tried to explain to him why we had thought it essential to release news of the incident, he walked backwards out of the room, barely able to speak." The British general continues: "I did not go after him. Having spent the past five months doing all I could to make the Coalition work, I was not going to risk splitting it now, in the middle of the battle." Nevertheless, he knew that "the USAF were not in line with our interpretation of events."

There are many unanswered questions concerning this blue-on-blue tragedy, and neither the British nor the American authorities have been prepared to help solve them. There is an obvious conflict of evidence between the official report issued by the pilots and the evidence given by the British air controller and the troops on the ground. Of the three men who knew the truth at the time of the incident, the British air controller and the two American pilots, the former has claimed that before being contacted by the two A–10s, he had already given target grid references to American planes that had successfully attacked Iraqi armor some twenty kilometers to the east of the British Warrior APCs. He asserts that he gave the same grid references to the two American pilots involved in the blue-on-blue, something that they deny. On the contrary, they claim that he gave them a clearance to attack anything within ten kilometers by saying that there were no friendly forces within that area. The pilots claim they selected their targets on the basis of identification. The air controller told the British Board of Inquiry that he had also issued a code word with the grid reference to reassure the American pilots that they were not being fed information by Iraqi intelligence, a precaution that may or may not have been necessary in view of the unsophisticated Iraqi defense system, but one that had been agreed at the outset of the fighting by Coalition members. Curiously, the two pilots deny receiving such a code word and furthermore claim not to know its significance. This conflict of evidence is so wide that one is left with no other conclusion than that someone is lying as part of a cover-up. Without full knowledge of the conversation between the pilots and the air controller, we cannot tell at this stage what really happened. However, even if the air controller gave neither code word nor grid reference and merely vaguely mentioned that there were no friendly forces within ten kilometers, how could the pilots have misidentified the Warrior APCs as T54/55 tanks and have claimed that they carried no markings? Apparently, once the A–10s had carried out their attack on the Warriors, they informed the air controller of the grid reference, which showed him for the first time that they had attacked a friendly force. He ordered an American reconnaissance plane to overfly the area and received the significant report that fluorescent air recognition panels could be seen from 6,000 feet and the type of vehicles could be identified from 14,000 feet. This report would seem to demolish the claims by American commanders that dust and sand could have obscured identification signs and perhaps even made the shapes of the vehicles difficult to identify. What is possible, of course, is that some of the Warrior APCs could have become dirtied by sand or dust, and some others perhaps lost their fluorescent panels in the wind, but thirty-seven vehicles in close proximity do not all suffer identical mishaps and transform themselves in shape into T54/55 tanks.

In public the American air commanders, notably Lieutenant General Chuck Horner, have been adamant that the error leading to the blue-on-blue was made by the British air controller. They claim that his comment that there were no friendly forces within ten kilometers gave the A–10 pilots carte blanche to attack anything they found within that area. Yet, even if the British air controller is lying when he asserts that he gave the pilots a grid reference, Horner is on record as saying that pilots had been instructed: "If in doubt, don't drop." How was it possible for the two pilots to fail to identify any of the hundreds of identifying panels and signs on the thirty-seven Warriors or notice the many other vehicles of the Third Battalion of the Seventh British Armored Brigade spread out across a perfectly flat plain? The desert was swarming with British vehicles, all thoroughly equipped with colored panels and inverted *V*s, as was apparent from television film shown only minutes after the attack.

The blue-on-blue of February 26, 1991, came as a shattering blow to a complacent British public, accustomed to seeing American smart weapons finding their way into the buildings in central Baghdad with all the skills of a cat burglar. The dangers that British and other Coalition troops were facing seemed far less than those that had been ever-present in the short, sharp Falklands War. The news that nine British soldiers, all young men aged between seventeen and twenty-one years, had been killed by a mistake seemed particularly shocking as it had been so unnecessary. "Friendly fire" was a new term in daily newspapers and on television reports. Old soldiers and those still serving knew it for what it was, an inevitable part of service life ranging from an accidental killing when a rifle was being cleaned right up to the misdirected bombing raid on Allied troops in Normandy in 1944, known as Operation Cobra. But to a public accustomed mainly to the losses in peacekeeping operations, the loss of nine soldiers to an attack by American planes was something that could not be ignored. Old prejudices surfaced and newspaper articles appeared, dredging up previous examples of American "gung ho" behavior in World War II or Korea. Author Denis MacShane, in the most complete survey of the incident in the Iraqi desert, enumerated characteristics of American pilots that might have contributed to the disaster. According to MacShane, the case of another blue-on-blue involving a senior American officer, Lieutenant Colonel Ralph Hayes, might offer parallels. Apparently, Hayes was flying an Apache helicopter that destroyed two U.S. Bradley APCs, killing two men the day after the British blue-on-blue. Hayes, on leaving the U.S. Army, gave a detailed interview on American television in which he attempted to vindicate himself. Human error clearly played a major part in his incident, as he misread the grid reference while in contact with unidentified vehicles at night. The result was that he believed himself to be over an Iraqi-held area and was immediately ordered by his ground controller to destroy the targets. But something worried Hayes at the time and he felt uneasy about attacking with missiles. He tried to strafe the mystery vehicles but his gun jammed. His uncertainty was manifest as he told the controller that he was finding it hard to pull the trigger. However, the controller was adamant and continued to order him to "Take 'em out." In response, he fired first one missile and then another, and two American lives were lost. The controller then realized that a mistake had occurred and informed Hayes that he had just hit American vehicles. In spite of state-of-the-art technology, a senior American pilot misread a number, and from that moment on, no amount of technology could save the lives of those two soldiers.

According to MacShane, the Hayes case reveals that audio and video recordings are kept of conversations between air controllers and pilots, as are the views of targets as seen by the aircraft's weapons system. Thus the conflict of evidence between the two A–10 pilots and the British air controller could easily be rectified by the production of this material. That it has not been produced in public indicates that a cloak of secrecy has been drawn over this particular incident that was not employed in the Hayes affair. If the British Ministry of Defense has seen the videos and heard the recordings, then it has entered into the American cover-up. If not, the Americans are operating a cover-up on their own behalf. Certainly, the authorities in the United States have gone to some lengths to conceal details of friendly fire incidents in the Gulf War, of which there were many. At least thirty-five U.S. servicemen died as a result of blue-on-blue incidents in the Gulf, and it is significant that in casualty reports the phrase "hit by friendly fire" has been deleted and replaced at high level by a handwritten "vehicle hit by enemy fire." Where U.S. soldiers have tried to speak of their experiences of friendly fire, they have been threatened with court-martial, according to reports in the *Washington Post.* In fact, it was not until August 1991 that all the families of the thirty-five American victims of friendly fire had been officially informed of the cause of their deaths. Truth may be the first victim in wartime, but there is no reason that truth should continue to be denied once the war has ended. The parents of the nine British servicemen who died deserve to know the truth of the events that led to the deaths of their sons. Nothing is gained by cover-ups except delay; the truth cannot be permanently suppressed.

In the course of the fighting in the Gulf, British soldiers were casualties in three further incidents that involved friendly fire. Though far less damaging than that involving the nine young men in the Warrior APCs, these cases are further proof of the increasing incidence of blue-on-blue episodes. The first of them occurred on February 26 at 1100 hours. An officer of the First Staffordshire Regiment, who had just alighted from a Warrior APC, was hit by shrapnel during a brief engagement with a Challenger tank of the Royal Scots Dragoons. The Challenger had opened fire, in poor visibility, on a number of Iraqi tanks, which turned out to be abandoned. The tank then fired at a group of Warriors from the Staffords, and although no damage was done, the officer was evacuated for hospital treatment. The poor visibility due to the dust storm was clearly the cause of the incident.

The following day, again at 1100 hours, two members of the Queens Royal Irish Hussars, traveling in Scorpion armored reconnaissance vehicles, fell foul of an American M1 Abrams tank, which opened fire on them although both Scorpions carried the normal Coalition markings and visibility was good. The cause of the confusion may have been that the Scorpions had both stopped to take the surrender of Iraqi troops, and this may have given the American tank crew the impression that they were, in fact, Iraqi vehicles. In any case, the M1 opened fire at a range of fifteen hundred meters and hit the first Scorpion in the front, injuring one soldier. The other Scorpion was then hit by machine-gun rounds and shrapnel from a main turret round. The turret gunner in the Scorpion was injured by the shrapnel. On discovering their error, the crew of the Abrams helped evacuate the wounded to a hospital.

The third incident occurred on the same day, but during the afternoon. Two soldiers from the Tenth Air Defense Battery of the Royal Artillery were wounded when their Spartan armored vehicles were fired on by British Challenger tanks. The tanks were using thermal sights to aid targeting, and this prevented the identifying signs on the Spartans from being seen. As a result, one of the Spartans burst into flames on being hit, but fortunately it was empty at the time. The other Spartan was damaged, but less seriously.

American losses from friendly fire were much greater than those suffered by British or other Coalition forces. In fact, almost twenty-five percent of American casualties were self-inflicted. Yet the case of the nine British soldiers killed by the mysterious blue-on-blue, mysterious because at least one government is unprepared to release the full facts, has come to symbolize the entire subject for many people. Chuck Horner's reaction to the bereaved parents' attempt to get at the truth—that it was "picking at a scab"—is unhelpful because even if the scab is a sign of healing, normality only returns when the scab is gone and forgotten.

1

Ground Warfare From Ancient Times to the First World War

ANCIENT AND MEDIEVAL WARFARE

The wars of antiquity and of the medieval period rarely speak of friendly casualties. Accidents on the battlefield must have been so common that they were scarcely worthy of comment unless they related to a leader of note struck down by mischance. It might be thought that the hand-to-hand nature of battle in these periods would preclude the sort of problems that have bedeviled more recent conflicts, and to the extent that missile fire—from bowmen, javelin throwers, and even primitive cannon—was directed at a seen target rather than indirect, this could be said to be true. Yet the use of pointed and edged weapons in close order was bound to offer a danger to colleagues as well as to enemies. In his book *The Western Way of War,* Victor Davis Hanson has shown that the ancient Greek battlefield was a place of "confusion, misdirection and mob violence", with "no clear demarcation between sides" and it was sometimes almost impossible to identify friends and allies in the chaotic fighting. Because uniforms were not generally worn in European warfare until the middle of the seventeenth century the opportunities for mistakes were great. As Hanson shows, the Greek hoplites from various states were almost indistinguishable from one another. One hardly had time to ask a man's allegiance before falling to blows with him, nor could one be certain that he was telling the truth. Thus,

> hoplite battle was frequently enough a free-for-all of sorts between infantry who looked, dressed, fought and spoke alike, and there is little wonder that often men had no idea whom they were fighting once their phalanxes had crashed and merged. Since the pressure of advancing and retreating lines varied, men might find their own colleagues out in front and nearly in their faces, while pockets of the enemy were at their side and perhaps already to their rear.

Thucydides described the situation that ensued during the Athenian night attack at Epipolae in Syracuse in 413 B.C.:

> The Athenians now fell into great disorder and perplexity . . . seeking for one another, taking all in front of them for enemies, even although they might be some of their now flying friends . . . They ended by coming into collision with each other in many parts of the field, friends with friends, and citizens with citizens, and not only terrified one another, but even came to blows and could only be parted with difficulty.[1]

This chaos, even more typical of a medieval mêlée, was naturally productive of much amicide. What is apparent, however, is that neither ancient warriors nor medieval soldiers thought fit to complain about this additional threat to their lives, reinforcing the view that it was seen as simply an unavoidable part of a soldier's lot. In several other ancient battles—notably at Cannae in 216 B.C. and Adrianople in A.D. 378—the troops of one army became so compressed by the enemy that they could do little other than strike at their own friends and inflict casualties on their own side. Ammianus Marcellinus, writing of the terrible defeat of the Romans under Valens by Fridigern's Goths at Adrianople, described how "the different companies became so huddled together that hardly anyone could pull out his sword, or draw back his arm, and because of clouds of dust the heavens could no longer be seen, and echoed with frightful cries." So confused was the fighting that many Romans fell victim to their own men at the rear, and nobody could escape until the frenzied slaughter had thinned the ranks and made more space for men to fight and run.

Thucydides describes the chaos that ensued within the Athenian army during the Battle of Delion in 424 B.C. against the Boeotians and their Thespian allies. The Athenians achieved a kind of pre-Cannae encirclement of the Thespians, but when the left wing of the Athenians eventually joined up with the right wing, the hoplites failed to identify each other as Athenians and fighting broke out. Hanson has suggested that the reason for these friendly casualties was that the lust for killing blinded the Athenian soldiers to everything but the so-called enemies around them. Inherent in this is a positional problem. The Athenians assumed that anyone in front of their battle line must be an enemy and therefore must be engaged and killed. In the confusion of hand-to-hand battle, this simple assurance was frequently wrong. Nevertheless, where warriors were dressed alike, spoke the same language, and were not obviously identifiable, position was the best assurance that the men you were fighting were the enemies you were supposed to be fighting. In the event of the two armies becoming intermingled, mistakes were certain.

Besides the positional problems inherent in hand-to-hand fighting in an era before uniforms were generally worn, Hanson has also drawn attention to the propensity for friendly casualties arising from the use of certain pointed or edged weapons. Accidental wounding or killing was an ever-present problem of service in the Greek phalanxes. And what was so in the Greece of the fifth and fourth centuries B.C. must also have been true of the phalanx formations used by the Swiss and the German landsknechts of the fifteenth and sixteenth centuries. The Macedonian sarissa reached the extraordinary length of 6.3 meters, according to Polybius, compared to the 5.5 meters of the Swiss medieval pike. This meant that when the Greek phalanx advanced into battle the sarissas of the first five ranks projected beyond the foremost men. The problem was that the sarissa was pointed at both ends, so that when the front ranks drew back to strike the enemy, they frequently wounded or killed men farther back in the phalanx. As Hanson reasonably explains: "After all, front line troops are fighting for their lives with leveled spears against the enemy. They could hardly worry about the danger of their own butt spikes to the men behind. At the same time, the men in the front row or two might themselves be wounded on the flank from the spear tips of those in the rank behind." An additional problem was tripping and trampling, which must have been a common problem in the confused fighting at the front of the phalanx, while if the front rank fighters were forced backward, they were likely to be impaled on the spears of the men behind them. Although the Roman legionary formations were quite unlike the Greek phalanxes, notably in their flexibility, several times in Rome's history its armies were trapped and so compressed as to cause a panic resulting in many casualties from friendly blows and trampling, notably at Lake Trasimene in 217 B.C. and at Cannae the following year.

The almost universal adoption of a butt-spike to the long, thrusting spear was not only an important technological addition to the range of weapons available in the battle between the phalanxes in ancient Greece, but a major factor in amicide. The spike was generally squared in shape and was made of bronze, about eight inches in length. Not only did it act as a counterweight to the spearhead, allowing better balance for the hoplites or later phalangites, but it also enabled the spear to be reversed in the event that the spearhead was broken off during the first shattering impact of phalanxes. As we have seen, this butt-spike undoubtedly inflicted many friendly casualties, but it had a sound military purpose. As the phalanx passed over fallen enemies, it was relatively easy for the hoplites to administer the coup de grace by a vertical thrust with the butt-spike. There is much evidence in the shape of squared holes in Grecian armor to suggest that this was a normal way to dispatch men on the ground.

Several times in the Middle Ages, perhaps at Morgarten in 1315, Roosebeke in 1382, and Morat in 1476, were scenes of slaughter and amicide such as occurred at Epipolae, Cannae, and Adrianople seen again. On a misty day at Roosebeke, a large force of Flemish rebels, commanded by Philip van Arteveld, drew up in defensive formation behind a ditch and a barricade of bushes, awaiting attack by a French army under King Charles VI and the constable of France, Olivier de Clisson. When the French did not immediately attack, the Flemings lost patience and moved out of their position in an attempt to reach a nearby hill. This maneuver proved fatal to an essentially infantry army facing one strong in heavy cavalry. As the mist rose, the Flemings were exposed on the march, and the French cavalry advanced on both flanks and encircled them. The Flemish pikemen were now hemmed so close together that they could not even reach the French to strike at them. In the press, many Flemings fell and were trampled under foot or suffocated, Philip himself dying in this way. The chronicles tell us that the Flemings were "slain by heaps, one upon another."

As John Keegan has demonstrated in his account of the Battle of Agincourt in *The Face of Battle*, during the infantry struggle, the simple art of keeping one's feet becomes a matter of life and death. In the tight scrimmage that was a feature of most ancient and medieval battles, death came quickest to the man who fell and left himself open to be trampled, suffocated, or penetrated by a stabbing blow from above. Many an apparently invulnerable knight, clothed from head to toe in armor, died when a lowly archer opened his visor and stabbed him in the face.

Keegan stresses the importance of the press of sheer numbers that was a feature of the battle as the dismounted French men-at-arms came to grips with their English equivalents. Although the French mass was far larger than the English, it did not necessarily increase their chances. In Keegan's words, "The unrelenting pressure from the rear on the backs of those in the line of battle [drove] them steadily into the weapon-strokes of the English [and denied] them the freedom for individual maneuver which is essential if men are to defend themselves." The Frenchmen at the front were thus reduced to helpless impotence, denied room to swing their weapons or even raise their arms. It was all they could do to keep their feet without falling. As Keegan explains:

> At Agincourt, where the man-at-arms bore lance, sword, dagger, mace or battleaxe, his ability to kill or wound was restricted to the circle centred on his own body, within which his reach allowed him to club, slash or stab. Prevented by the throng at their backs from dodging, side-stepping or retreating from the

> blows and thrusts directed at them by their English opponents, the individual French men-at-arms must shortly have begun to lose their man-to-man fights, collecting blows to the head or limbs which, even through armor, were sufficiently bruising or stunning to make them drop their weapons or lose their balance or footing. Within minutes, perhaps seconds, of hand-to-hand fighting being joined, some of them would have fallen, their bodies lying at the feet of their comrades, further impeding the movement of individuals and thus offering an obstacle to the advance of the whole column.[2]

Ironically, the only person we know by name who died by trampling and suffocation was on the English side, the Duke of York, yet hundreds of French knights undoubtedly died in this fashion unnoticed, trod under foot by their own side. As Keegan demonstrates, once the front men fell, those behind would have to try to step over them, for to step on them risked tripping yourself. "Once the French column had become stationary, its front impeded by fallen bodies and its ranks animated by heavy pressure from the rear, the 'tumbling effect' along its forward edge would have become cumulative."

The hacking and slashing weapons used by the men-at-arms in medieval battles must have inflicted many injuries on their own colleagues. There simply was not the space in a medieval mêlée for the use of the Swiss halberd or the English bill without risks to colleagues alongside or behind. This must have been a factor taken into account by medieval commanders, who concluded that the offensive capability of such weapons more than compensated for the friendly casualties they caused. At Hastings in 1066 and presumably at Stamford Bridge in the same year, where there were axe-men on both sides, the English housecarles fighting left-handed and swinging huge battle-axes with five-foot helves cannot fail to have struck friend as well as foe. Although these men were well-trained professional warriors, space alone would have determined how easily they could have swung a weapon requiring such freedom of action. To make matters more difficult for the housecarles, they had been positioned at the front of the English line on the hill at Senlac and had stacked up behind them the mass of the Fyrd, four to five thousand men who ranged in quality from the professional housecarles of the numerous English thegns to excitable shire levies, farmworkers armed with flails, spades, hoes, and pitchforks. The chroniclers tell us that as they advanced up the slopes toward the English position, the Norman cavalry and infantry were struck by a hail of missiles, hurled by the fyrdmen. It must have been scarcely safer or more palatable to have stood with your back to the Fyrd, as did the housecarles, while these missiles, ranging from stones to throwing axes, flew overhead, than to be the Normans facing the barrage. That men fell on both sides from the blows and missiles of their comrades was inevitable in such close quarter fighting.

Missile misdirection must have been a cause of many friendly casualties in the days before firearms became generally employed. Massed archery fire, which was a feature of all English battles of the fourteenth and fifteenth centuries, would have always involved a proportion of accidental killings. While the two armies were well-separated, the archers could have fired with relative impunity, but when the main bodies of men-at-arms became heavily engaged, as against the French at Crécy in 1346, Poitiers in 1356, and Agincourt in 1415, the archers would have had to cease firing at a time when friendly casualties exceeded an acceptable level. At that moment, the archer became militarily helpless, unless he was able to turn his hand to the kind of killing that was a feature of the battle at Agincourt, where they took on the task of dispatching fallen knights, who found it difficult to get to their feet in the muddy and slippery conditions, wearing sixty pounds of armor. Nevertheless, as lightly armed and unarmored infantrymen, they would have kept firing as long as possible, thereby risking friendly casualties, rather than lose their only advantage over the armored pikeman or sword and buckler man-at-arms.

Missile misdirection was a feature of the Battle of Towton in 1461, during the Wars of the Roses. Here high winds driving snow into the faces of the Lancastrians created a visibility problem that was exploited by Lord Fauconburg, one of the Yorkist leaders, who ordered his archers to fire a cloud of arrows "indirectly" toward the unseen enemy, so that the strong wind carried them far beyond their normal range. Convinced that the Yorkists were well within range—though hidden by the blizzard—the Lancastrian archers fired back into the wind, only for their arrows to fall harmlessly short or more dangerously onto the heads of their own forward troops.

A second and even more serious visibility problem occurred at the Battle of Barnet in 1471. This time the problem was fog. As we have seen, the lack of uniforms created serious problems of identification, and consequently contributed to accidents on the battlefield. During the Wars of the Roses, many men at arms and soldiers wore the colors or liveries of their lords, and these emblems or devices helped identify friend and foe, even in the heat of battle.

At Barnet, misidentification had profound consequences. The problem arose because of the similarity, or so it might appear in the fog of war—literal and physical—between the banner of the Earl of Oxford, commanding the right wing of the Lancastrian army, and that of the Yorkist king, Edward IV. In the murk, it was possible to mistake the radiant star of the de Veres (Oxford's standard) for Edward's device of a brilliant sun with rays. In battle, men are frequently governed less by their hopes than by their fears, and so as Oxford brought up his wing to the rescue of the Lancastrian center, under the Duke of Somerset, instead of their thanks, he received a volley of arrows from a flank guard of their archers. His cavalry was thrown into confusion and began to shout, "Treason! Treason!" to add to the confusion. With the enemy in front and a host of cavalry emerging from behind them bearing the devices of what looked like the Yorkist blazing sun, Somerset's men assumed the worst and panicked. As the Lancastrian center collapsed, Somerset's men fought tooth and nail against the troops Oxford had brought to their succor, while the latter, convinced that they were being attacked by men who had changed sides and become traitors, fought back equally vigorously. The beneficiaries, Edward's Yorkists, leaving Somerset and Oxford to their internecine struggle, were left with the relatively simple job of routing Warwick's remaining reserve and winning a decisive victory.

The most celebrated example of a medieval blue-on-blue must certainly be the destruction of the Genoese crossbowmen at the Battle of Crécy by their French paymasters in 1346. As the French army advanced in disorder toward the English,

King Philip VI ordered his crossbowmen, six thousand or so Genoese mercenaries under the command of Odone Doria and Carlo Grimaldi, to open the attack. But the Genoese were exhausted after marching for nearly eighteen hours, and when ordered to fire, they complained that their bowstrings had been soaked by the intermittent rain. But the king was insistent, and so the Genoese marched ahead of the French army and prepared to open fire on the silent ranks of the English, drawn up on a slight hill before them. The first volley from the Genoese fell short, and they never managed a second because they were struck by an astonishing arrow storm of perhaps sixty thousand arrows in the next sixty seconds. The Genoese panicked and tried to get out of range of the deadly English longbows, but they could not break through the serried ranks of the French cavalry. The chronicler Froissart continues the story to its ridiculous denouement: "The French had a large body of men-at-arms on horseback to support the Genoese, and the king, seeing them thus fall back, cried out, 'Kill me those scoundrels, for they stop up our road without any reason.' "[3]

The result was that lines of French knights commanded by the Duc d'Alençon charged straight into the Genoese ranks and tried to trample them underfoot. In the words of Sir Charles Oman: "This mad attempt to ride down their own infantry was fatal to the front line of the French chivalry. In spite of themselves they were brought to a stand at the foot of the slope, where the whole mass of horse and foot rocked helplessly to and fro under a constant hail of arrows from the English archery."

As was seen at Agincourt and elsewhere, the cutting and slashing weapons prevalent during the Middle Ages were a major source of amicide, in view of the hand-to-hand nature of much medieval warfare. By the fifteenth century, the weapons wielded by men-at-arms had reached a limit, imposed by the capacity of individual warriors to wield them. The halberd of the Swiss was a deadly weapon, but employing it required too much space. In addition, its length of six to seven feet was inadequate to keep enemy cavalry out of killing distance. Its replacement by the pike—eighteen to twenty feet in length—imposed on the Swiss infantry a new and even more successful adaptation of the ancient phalanx formation. But the halberd was not totally rejected, and when the pike-phalanx failed to make the initial breakthrough, halberdiers from the center or rear of the phalanx emerged, to try to cut their way through the enemy pikes. To counter the effectiveness of the Swiss pikemen of the fifteenth century, the Habsburg emperor, Maximilian I, created his own bands of German pikemen, known as the Landsknechts. These men, who often fought against the Swiss as mercenaries, attempted to challenge the Swiss in the "push of pike" and wielded weapons of an even more deadly nature, like the huge six-foot-long double-handed sword, which must have endangered friendly forces as much as the enemy, or the halberd, the voulge, the glaive, the partisan, or the spetum, all huge slashing and crushing weapons. In the hands of the Doppelsöldner, the double-handed sword, known as the Zweihänder, must have been a fearsome thing, but limited by the space necessary to wield it. In the hand-to-hand fighting that was a feature of fifteenth- and sixteenth-century battles, and before the advent of effective firearms, it is inconceivable that such weapons could have been wielded safely without a significant number of friendly casualties.

Blaise de Monluc, that grim and doughty professional warrior, writes of a friendly fire incident in which he was involved during the Habsburg-Valois Wars of the early sixteenth century. During a Spanish siege of Marseilles in 1536, Monluc and a company of men had raided the flour mills at Auriol on which the Spaniards were dependent. But in trying to return to Marseilles, he was mistaken for a Spaniard and brought under friendly fire:

> Coming near to Notre Dame de la Garde, the captain of the castle taking us for the enemy let fly three or four pieces of cannon at us, which forced us to shift behind the rocks. From thence we made signs with our hats, but for all that he ceased not to shoot till in the end, having sent out a soldier to make a sign, so soon as he understood who we were he gave over shooting . . . [4]

The introduction of early cannons and handguns greatly increased the incidence of friendly fire. The simple factor of unreliability meant that numerous accidents became a feature of battles and sieges. The untimely death of King James II of Scotland in 1460, during the siege of Roxburgh Castle, was a pointer to the future. James was personally inspecting his pride and joy, a large hooped bombard to which he had given the name Lion. When he tried to fire it, it exploded, showering pieces of wrought iron in all directions. A large piece of casing hit the king in the chest and killed him instantly, while other pieces wounded the Earl of Angus and several gunners standing nearby. The fifteenth century was the age of the great cannons, built to satisfy the wildest fantasies of conquerors like Charles the Bold of Burgundy or the Sultan Mahomet II, who used the greatest artillery train in the world at that time to batter the walls of Constantinople in 1453. But the lifespan of these monster guns was very short, and they exacted a high price for their services in dead and mutilated gunners.

THE EARLY MODERN BATTLEFIELD

Friendly fire was no less a feature of seventeenth- and eighteenth-century battlefields. The infantry of the day fought in lines, usually three or four ranks deep. The problem of maintaining effective musket fire in such deep formations was such that in the latter half of the century, the three-rank formation, first adopted by Frederick the Great of Prussia, became the pattern for Austrian, Russian, and French armies. Even so, the danger for men in the front two ranks of having their heads blown off by the fire of the third rank men was considerable. Perhaps the most extreme example of this occurred in the English Civil War, during the siege of Basing House in 1643. The house was a powerfully fortified building and was heavily garrisoned by Royalist troops. Parliamentary regulars, supported by men from the London Trained Bands, were besieging the south face of the building when Lieutenant Archer brought up the men of the Tower Hamlets Regiment, the Green Auxiliaries of London, and the redcoated Westminster troops. In spite of their colorful appearance and their impressive names, these men were closer to being militiamen than regulars and were poorly trained, with little experience on the battlefield. Once drawn up, the three lines of infantry forgot their drill completely, and there followed a most "lamentable spectacle." The front rank had been trained to fire a volley and then retire behind the other two ranks to reload and fire again when their turn came. Instead, in the excitement—or stress—of battle, all three ranks fired at the same time, and the front rank was completely wiped out by the fire of the rear-rank men. Over seventy Parliamentarians were shot down in an instant by their fellows, one of their officers wryly commenting, "Some on both sides did well, others did ill and deserved to be hanged."

Charles Carlton, in his book *Going to the Wars*, gives a vivid picture of the experience of the English civil wars between 1638 and 1651 for both the soldiers and the general population of the British Isles. The standard of military training in England under the early Stuarts was very low indeed, and English armies were generally recruited from the worst possible material. While planning the expedition to Cadiz in 1625, the Duke of Buckingham raised an army of ten thousand men. But these were no better than an undisciplined rabble of pressed men, the dregs of society, untrained, unfed, and badly clothed. One of his senior officers told the duke that "the army, both by land and sea, [was] in a very miserable condition for want of clothes," but Buckingham paid little attention. For months, the soldiers were not paid, and not surprisingly, many deserted or fell sick. In August 1625, a new impressment of two thousand men were billeted on the farms of South Devon, and trouble immediately broke out when the farmers discovered that their guests had no money to pay for their food. The destitute soldiers took the law into their own hands, roaming the countryside looking for sustenance, killing sheep, and threatening violence. When a senior officer reviewed the new recruits, he found that of the original two thousand, two hundred were quite unfit for service through ill health: Twenty six of them were aged over sixty, four men were blind, one was raving mad and a number of others deformed or severely maimed. The corruption of the impressment system produced many anomalies, not least of which was sending one man who had no toes and another who had one leg only half the length of the other. It is with no surprise that one reads the unfortunate history of the expedition. It was impossible for the commander, Viscount Wimbledon, to weld such poor manpower into a proper military force. Once in Spain, the English soldiers became so drunk that they degenerated into a raging mob, who besieged Wimbledon in his own quarters when he tried to stop them drinking the Spanish wine, shooting each other in their madness. Eventually, Wimbledon's personal bodyguard had to shoot some of their own colleagues to save the commander. When they were finally engaged in battle with the Spaniards, Wimbledon wrote, "They made few or no shot to any purpose, blew up their powder, fled out of their order, and would hardly be persuaded to stand from a shameful flight . . . the landmen were so ill exercised that when we came to employ them, they proved rather a danger to us than a strength, killing more of our own men than they did of the enemy."

Spared the horrors of the Thirty Years War in Germany and Central Europe, the English saw no need to professionalize their army, and as a result, in the early stages of their own civil war, incidents of military ineptitude, including friendly fire, were legion. Carlton records that during the First Bishop's War against the Scots, the English troops were equipped with faulty muskets, some with no touch-holes or with broken butts, poorly glued together. In a way, the men got the weapons they deserved. One man accidentally fired his during the march north under the Earl of Holland and killed a gentleman's son riding nearby. Other soldiers, in their drunkenness, fired bullets through their own officer's tents. On one occasion, during the civil war, an unruly soldier accidentally shot a bullet through the king's own tent. As early as July 1642, the records show a soldier shot in the foot by his own gun and another shot in the back by a rear-rank man. Even in the best hands muskets were often unreliable and exacted a high cost in lives among those who used them. Lieutenant Colonel Arthur Swayne was "slain by his boy, teaching him to use his arms. He bid the boy aim at him (thinking the gun had not been charged) which he did only too well." Captain John Francis, we are told, forgetting that his piece was loaded, "shot a maid through the head and she immediately died." During an argument with two Oxford undergraduates, Captain Stagger's musket went off accidentally, killing a woman who was shopping at a stall next door. A band of rowdy Royalist musketeers under the command of Sir Ralph Hopton were, according to Charles Carlton, responsible for two accidental killings in the space of a few days. In the first instance, a gentleman volunteer by the name of Christopher Berry was killed by "the going off of a musket unawares," while two days later one Thomas Hollomor was killed in the same way. In February 1645, we are told that Thomas Hills was killed by his own musket, and Lieutenant Vernon was killed when two bullets from a pistol were accidentally discharged by Captain Gibbon's groom. A Royalist officer confirmed the amount of accidental damage caused by the seventeenth-century firearms when he commented, "We bury more toes and fingers than we do men." Misfires, it has been estimated, could reach as high as 18 percent. Even the doughty Parliamentarian John Hampden succumbed to his own overloaded pistol at the Battle of Chalgrove Field.

At the siege of Farnham Church in November 1642, Parliamentary pikemen trying to force their way into the building became so congested in the church graveyard that a number of them were impaled on the pikes of their rear-rank comrades. Yet if problems in raising good pikemen and musketeers were great, getting skilled gunners and artillerists was of an entirely higher level of difficulty. And allowing unskilled louts to operate heavy guns and barrels of powder was just asking for trouble. Some of the first casualties of the civil war were the six Royalist gunners killed when their gun burst during an early skirmish. At the siege of Limerick in 1642, Master Gunner Beech, unwisely relying on his unskilled assistants, died when they overloaded a cannon with powder. When Beech fired the gun, the barrel exploded, blowing him and—one hopes—the culprits to Kingdom Come. Accidental explosions were always a danger in the age of gunpowder. Eighty barrels of powder had been stored in Torrington Church during the English civil war, and their accidental explosion killed more than two hundred of the garrison guarding the town. It was not always possible to blame a malign fate for such a fiery intervention—sometimes it was just plain stupidity. The Earl of Haddington and some of his staff officers were blown up when they called for candles, and one of their servants set them up on a nearby barrel of gunpowder. One powder manufacturer, Edward Morton, died with his entire household, except his wife, while mixing powder for the king. Other accidental friendly fire involved the foolish Captain James Hurcus, "hoist by his own petard" when, during the siege of Gloucester, he climbed out of his trench to see if the grenade he had just thrown had exploded. It had not, but it did once he retrieved it.

The literal fog of battle in the age of pike and musket was very severe. In a large battle of the English civil war, such as Marston Moor in 1644, hundreds of cannons and perhaps twenty thousand men firing muskets at the same time produced great billows of sulphuric smoke, reducing visibility for everyone on the field. In such conditions, mistakes through misidentification were inevitable. And because identifiable uniforms did not become common until the mid-seventeenth century the chances of friendly fire incidents were very high. In May 1645, at the small Battle of Crewkerne, Royalists under General Goring fought for nearly two hours against each other, entirely mistaking their colleagues for Parliamentarians. In place of uniforms, identification was limited to battle signs—sometimes colored cloth or heather—but these were so easily acquired and discarded that they were worse than useless. The Parliamentarian general, Sir Thomas Fairfax, actually rode through the entire Royalist army at Marston Moor without being recognized by simply removing the white cloth that the Parliamentarians had been wearing in their hats for identification. Slipping it in a pocket, he escaped capture and simply replaced it as he regained friendly lines. This worked well as long as you remembered which side was wearing what. At Edgehill in 1642, Sir Faithful Fortescue—belying his name—deserted the Parliamentary army and led his troop of soldiers over to the king's side. But, failing to remove their "orange tawny scarves", they were attacked by their new allies, and eighteen men were cut down and killed.

The closing stages of the Duke of Marlborough's victory at the Battle of Oudenarde on July 11, 1708, contained elements of such confusion between General Overkirk's Dutch troops and General Cadogan's British command that it was finally necessary to allow many of Marshal Vendome's trapped

French soldiers to escape in order to avoid a catastrophic blue-on-blue incident. Like the Athenians at Delion in 424 B.C., Marlborough's allied army was in the process of encircling their opponents, and the trouble began when the troops of each encircling flank began firing on each other. To add to the confusion, visibility was poor, as dusk was setting in rather early for midsummer and a light rain was beginning to fall. In the deepening gloom, Marlborough and Prince Eugene felt they had no alternative but to order a general cease-fire, as Overkirk's and Cadogan's troops were by now thoroughly intermingled, and mistakes were certain.

The only way through this maze of military incompetence was discipline and drill, and the eighteenth century was the great age of the drillmaster. The model for all European—and American—armies was the one established by the Prussian kings. But even Prussian-trained armies had difficulties in the heat of battle. While commanders saw advantages in having the strongest, tallest, and fiercest fighters in the front line—for bayonet and hand-to-hand struggles—the taller the front-line men, the greater the chance of their being shot by the shorter men in the rear rank. The way out of this apparent impasse was to drill the men so thoroughly that they worked with machinelike efficiency. At least that was the theory. Moreover, provided that the troops kept in close formation, the musket barrel of the third-rank men was supposed to project a few inches beyond the heads of the front-rank men, thus removing the danger of friendly casualties, except perhaps through deafness. However, although the first volley in a battle might have been fired in such a coordinated fashion, later volleys were fired in the confusion of battle. Once men began to fall from enemy musket or cannon fire, the formation became disordered, and when it was necessary to repel a cavalry charge, even the most well-drilled soldiers lost their sense of formation. It was at this point that mistakes were made. What had seemed so easy on the parade ground became unattainable even for the automatons of Frederick's Prussian army or the Austrians of Marshal Daun. Discussing poor fire discipline among eighteenth-century soldiers, Christopher Duffy writes:

> In the excitement and noise of action there was no guarantee that the soldier would notice that his musket had failed to fire and in such a case he loaded round after round until five, six or more charges were superimposed. If the first round now took fire the barrel exploded like a bangalore torpedo. The barrel was also liable to burst when the muzzle was obstructed by dirt or snow, when a bullet happened to stick in the bore or if the ramrod was left in the barrel after loading.[5]

At the Battle of Kolin in 1757, an Austrian officer noted that "many a brave lad fell dead of wounds inflicted from the back, without having turned tail to the enemy . . . The surgeons were later ordered to inspect the battlefield, and it transpired that these mortal wounds had been delivered by men of the rearward ranks, who carelessly mishandled their muskets in the heat of the fire." In fact, Kolin saw another outbreak of friendly fire, which Christopher Duffy has described as a "feu de joie." So delighted were the Austrian rank and file after defeating the Prussians and reversing a long string of defeats that many soldiers broke ranks and fired their muskets in celebration, careless of where their bullets went.

The disintegration of formal lines, which was a normal consequence of the chaos of battle, often followed after the first two or three volleys of the firefight. Even the Prussians sometimes succumbed to a "general blazing away," so that as the adrenaline flowed, men began to load faster and faster and fire with less thought of consequences. Whereas the front rank alone was often ordered to kneel when it fired, soon men at the rear began to kneel for safety and found their own comrades blocking their aim at the enemy. At Zorndorf in 1758, retreating Prussian troops ran straight into their own rear ranks taking heavy casualties from their own infantrymen. So great was the confusion that previously well-ordered units panicked and began to fire this way and that, inflicting more casualties than the pursuing Russians.

At the battles of Parma and Guastalla in 1734, the French and Austrian front-rank infantry both fired from a kneeling position. But, no doubt feeling safer close to the ground, many of them refused to stand up to reload, and it was observed that soldiers on both sides lay full length, firing "in the manner of the Croats'' from a prone position. With their aim obscured by the legs—if not the bodies—of many of their colleagues, they could not fail but to shoot many of them in the back. In the heat of battle, it may be suggested, many of the soldiers were scarcely aware of what they were doing.

The discipline thought essential in the eighteenth century was thus a two-edged weapon. Its positive virtue was to bring order to an essentially chaotic situation and give the individual soldier the certainty of being part of a much larger machine, in which he was merely a cog—and an unthinking one at that. It maximized firepower at the beginning of a battle but could do little to check the speedily developing chaos that was bound to take over as the battle progressed. Frederick the Great admitted as much to his friend Catt at the Battle of Zorndorf. He first asked Catt if he had understood what was going on during the battle, to which Catt replied that he could make no sense of the various movements. Frederick reassured him: ''You were not the only one, my dear friend. Console yourself, you were not the only one.'' Thus for any commander it was vital for the first two or three volleys to be administered with parade-ground precision, after which matters began to move into the realm of chance.

The obverse side of producing well-drilled automatons on the Prussian model was that, deprived of the power of initiative or flexibility, European armies of the eighteenth century were often at a disadvantage when fighting in unfamiliar conditions and against unconventional opponents. This had a significant bearing on the incidence of friendly fire, which was essentially a product of the disintegration of discipline in the infantry firefight. One of the best examples of this can be seen in General Edward Braddock's disastrous campaign on the Monongahela River in North America against the French and their Indian allies in 1755. Here the formalized maneuvers and rigid linear formations of the British troops were unsuccessful against the mobility and flexibility of the woodsmen, who used concealment and scattered alignments to reduce the effectiveness of enemy firepower and discipline. Nor could the officers, however brutal their application of punishment to their men, ever recapture in the forests of

North America the parade-ground qualities that were the essence of European warfare.

The two regiments with which Braddock marched to Fort Duquesne were made up of relatively poor material, good enough perhaps to exchange volleys with their French counterparts, but low on initiative and, what was far worse, low on morale. The Forty-fourth, under Colonel Sir Peter Halkett, and the Forty-eighth, under Colonel Thomas Dunbar, had seen no recent wartime service and were taken directly from their peacetime cantonments in Ireland. Both were to sail from Cork under strength—with just 340 men out of a usual strength of 700. In fact, the plan was for them to be made up to full strength on arrival in America, by recruiting from the colonial population. This was an unwise plan, as the new recruits would have no time to accustom themselves to European standards of discipline. In recognition of this, a last-minute decision was taken to increase the British element to five hundred in both the Forty-fourth and the Forty-eighth by calling in men from other regiments. The result was that Braddock's army became a dumping ground for the "undesirables" that other regiments did not want. By the time he led his force to Fort Duquesne, the cadre strength of the Forty-fourth and Forty-eighth had been dangerously weakened, with each containing nearly a third of inexperienced colonial recruits as well as a large number of misfits and difficult men from other British regiments. It was a formula for disaster and reflected the view that second-rate troops would be good enough to overcome French colonials and their Indian allies.

Before Braddock left England, he was given clear instructions by the commander in chief, the Duke of Cumberland, known far from affectionately as the "butcher" for the treatment of the Scots at Culloden in 1746 and afterward. Dealing with a martinet like Braddock, one might have felt that Cumberland's strictures about discipline were hardly needed, yet the duke was wise enough to warn him about the panic that Indians might cause to green and undisciplined troops, particularly in North American conditions. Yet it might have been wiser on Cumberland's part to ensure that Braddock's army did not contain such troops, notably the four hundred or so colonials, many of whom would break and run at the first shot, or the three hundred or more British undesirables who would serve only to lower the morale of the rest of the redcoats. It was asking too much of Braddock and his officers to turn such poor material into an efficient fighting force in a matter of months. And nowhere would this deficiency be more apparent than in the quality of their fire discipline, rendering the likelihood of substantial friendly casualties a near certainty.

Braddock left Fort Cumberland on June 10, 1755, to begin his march to Fort Duquesne. Progress was slow—just thirty miles in eight days—and the gloomy, interminable forest exercised a malign influence on the minds of the simple Irish lads who comprised much of the British regiments. Nor were the American recruits any happier. Most had been accustomed to life on the eastern seaboard, and they had no experience with the wild country of the Ohio. At each stage of the march, Braddock had maintained the tightest possible discipline, aware that there was always the chance of being ambushed by Indians. On July 8, he successfully crossed the Monongahela River at Turtle Creek and concluded that the greatest danger was past. Only a few miles separated the British force from the French in Fort Duquesne. But with the hardest part already behind him, Braddock seems to have relaxed his precautions against ambush at the very moment that the French and their Indian allies were about to strike. The British advance guard, commanded by Lieutenant Colonel Gage, was surprised by a group of Indians, led by a French officer, who rushed down the path toward them and then fanned out into the trees on either flank. As Gage formed up his men with great precision, aligning them to the front, the Indians opened up a galling fire from behind cover. Gage replied with a series of well-ordered volleys against a virtually invisible enemy. By chance, one British bullet killed the French commander, Captain Beaujeu, but that was virtually the last success of the day for the British redcoats. While Gage's men stood, facing the front, the Indians ran from tree to tree, working their way down both British flanks, giving the stationary redcoats the unpleasant feeling that they were being surrounded. Gage now fired his two cannons into the trees, splintering wood, bringing down branches and showers of leaves. The Indians, meanwhile, stayed behind cover and picked off the redcoats with ease. It was like no battle that the British officers and their men had ever seen or heard of. Discipline cracked, and soon the redcoats took to their heels, rushing in panic back to the main part of the column. To make matters worse, the flank guards abandoned their positions and joined the scramble into the center, allowing the Indians to enclose the British column on three sides.

The British position was very dangerous but not hopeless. Everything depended on how Braddock brought his main force into action. Gage had already made things difficult by losing control of the advance guard and allowing it to run pell-mell into the main body, spreading despondency and disorder. As one observer commented on the chaos of the British troops, they were "men without any form of order but that of a parcel of school boys coming out of school." The problem for Braddock was that his regular soldiers had been marching on either side of a long line of wagons. To reach the firing lines, the men had to march under fire and in the chaos of battle up the side of these obstructions until they could form up at the head of the column. With Gage's men rushing back toward them, confusion and panic were almost inevitable. Moreover, the Indians had picked off many of the officers, notable by their uniforms and gorgets, so that there was no one to give orders, and the men were not trained to act on their own initiative. No European army could be expected to act effectively in the event of an ambush, and so, thrown back on their own resources, the common soldiers tried to do two basic things: fire as soon as they could reload, and find cover, both things that were absolutely anathema to the eighteenth-century method of warfare. With masses of soldiers, in improvised ranks twelve deep, firing anywhere, the majority of casualties would be friendly ones. With an enemy concealed behind trees, the only targets were British ones, and hundreds of men fell with bullets in their backs fired by terrified colleagues. Deprived of the certainty that fire drill had given them and deaf to the entreaties of the few officers left standing, the British soldier could think of nothing but to keep firing his musket in an attempt to ward off a fearsome enemy he could not even see. As an eyewitness commented later, "The confusion and destruction was so great, that the men fired irregularly, one behind another, and by this way of proceeding many more of our men were killed by their own party than by the enemy, as appeared afterwards by the bullets that the surgeons extracted from the wounded, they being distinguished from the French and Indian bullets by their size."

Given the conditions and the fact that the enemy was already using cover to very good purpose, it might be supposed that the British soldiers would themselves find cover behind trees. But the idea of breaking formation was unacceptable to men drilled in the traditions of the Prussian system of Frederick the Great and his father, the "sergeant major" king, Frederick William I. Braddock went purple with rage when he saw redcoats hiding behind trees, beating them black and blue with the flat of his sword, forcing them back into the open, and calling them cowards. But as one of the men stoutly replied: "We would fight if we could see anybody to fight with." One Virginia captain ordered his company of colonials to take cover behind a fallen tree trunk only for them to be fired on by British regulars who mistook them for the French. When George Washington, one of the few officers—American or British—to survive the massacre, asked Braddock's permission to scatter three hundred men among the trees and fight the enemy Indian-style, the general lost all control and threatened to run him through with his sword. Just before Braddock was himself shot and mortally wounded, presumably by an enemy bullet, he is reputed to have shouted: "We'll sup tonight in Fort Duquesne or else in hell!" Nobody could doubt Braddock's courage, but even by the standards of the British army of the time, his performance under extreme pressure left much to be desired. His inflexibility was to be both his own undoing and that of his entire command. Discipline was one thing, but a closed mind was no adequate response to a crisis. Washington's suggestion, or indeed the advice of the Indian scout Scaroudy that an immediate retreat was a good idea, could have saved many lives and enabled the expedition to be reorganized for a second attempt on the fort. The worst response was simply to stand still in a clearing between the trees and allow skilled woodsmen to snipe officers and men alike at their leisure. Nor should the massing of men together, firing their muskets into the backs of their fellows, be confused with the kind of fire discipline expected of good European officers. Braddock had allowed himself to be ambushed in the first place, but what was worse was that the arrangement of his column—with too little distance between the various parts and with the main infantry body divided in half by the wagons and cannons—ensured that if he were surprised he would not have any time to form his men into drill lines, after which they would simply dissolve into a frightened rabble doing far more damage to themselves than to an unseen enemy. In such a case, the incidence of friendly fire would prove far heavier than on a modern battlefield, where each soldier fights as an individual as well as part of a larger unit. Ironically for the young Washington, some thirty-six years later, as president, he ordered an American army under General St. Clair into the forests against Indian opponents and suffered a similar disaster on the Wabash River. Both American regulars and militiamen succumbed to the fire of concealed Indian snipers and in return inflicted dozens of casualties on their own men. Few of the new recruits knew how to use their muskets properly and, according to one survivor, did much slaughter of the "twigs and leaves of distant trees," as well as of their fellow Americans. Like Gage's cannons, St. Clair's proved impotent, defoliating the forests but inflicting no harm on the Indians. The parallel between Braddock's and St. Clair's disasters was not lost on contemporaries, yet in the latter case fewer of the wounded survived long enough to have their wounds dressed and the offending bullets examined.

A combination of low morale, poor discipline, and a night march put paid to the Austrian army of the Emperor Joseph II near the village of Karansebes in Transylvania in 1788. In almost certainly the most remarkable example of amicide in all military history, the Austrian army disintegrated and fought itself, inflicting thousands of casualties in a series of panicky firefights in the darkness. And at no point during this grim and unnecessary slaughter did an enemy soldier come within miles of the battlefield.

The fiasco at Karansebes was, in a sense, a natural outcome of the fracturing of the tight discipline that was the most important part of eighteenth-century European warfare. Like their counterparts in Prussia, Russia, France, and England, the

Austrian infantry were drilled to a point where individual thought was unnecessary and initiative impossible. If the normal conditions of engagement were to change, as for example in a night action where the drill sergeants were unable to form the soldiers up in parade-ground order, or where an enemy exploited the conditions by hiding behind trees and refusing to form up in ranks to be shot down by well-ordered musket fire, soldiers were robbed of the certainty that drill had given them and thrown back on their own abilities, which had been beaten out of them in the interests of conformity.

In the case of the Austrian troops, they were also suffering from low morale brought on by a series of unexpected setbacks at the hands of the despised Turks, as well as poor health resulting from the emperor's absurd decision to camp in a malarial area near Belgrade. In the space of six months, 172,000 of his troops had fallen ill with malaria, of whom 33,000 had died. Nevertheless, in spite of the fact that morale was low and the condition of even the supposedly fit men was questionable, when news was received that a Turkish army, under the Grand Vizier, was approaching, the emperor marched off with just half of his army to force a battle.

The massed Austrian columns reached the town of Karansebes in good order, flanked on the march by regiments of Hussars. As night fell, the army crossed a bridge, watched by crowds of Wallachian peasants. Apparently some of the Hussars then stopped to buy liquor from the peddlers. However, when some of the infantry, tired from a day of footslogging, left their ranks to buy some wine as well, they were driven away by the Hussars. This was a job for subaltern officers, but the indiscipline was not stamped out and the disorder grew. The infantry, enraged at the arrogance of the cavalrymen, fired some shots in the air and tried to frighten them by yelling "Turci! Turci!" ("Turks! Turks!"), pretending they were about to be attacked. Joining in what seemed to be a bit of fun, the now drunken Hussars also shouted, "Turci!" and let off some shots as well.

What had started as a minor scuffle suddenly flared into a major incident. The rear columns of the army were still approaching the bridge and, hearing firing and shouts, and in the darkness perhaps expecting the worst, assumed they had been ambushed and began to panic. Men began firing at each other in the darkness. But when officers started to rush up and down the columns shouting "Halt!", the panic-stricken soldiers thought they were shouting "Allah!" It seemed that the whole Turkish army had fallen upon them in the darkness. Clearly, something disastrous had happened. Why else was there so much firing, and why were the officers shouting so desperately? Ahead, they could see the flashes of gunfire and all the turmoil of battle, with horsemen riding this way and that without any kind of order. The baggage handlers and transport workers at the rear of the army, afraid of being cut off by the ambush and slaughtered by the Turks, panicked and drove their wagons through the massed troops ahead, knocking soldiers in all directions and spilling many into the waters of the river. A great roar of terror was heard and thousands of men began to stampede in the darkness.

The emperor, himself sick, was traveling in an open carriage. The first he knew of the disaster was when he heard a sudden outburst of shouting as a flood of men, horses, and wagons swept by him, throwing his carriage off the road and tipping him into the river. Although he mounted his horse and with his staff officers tried to bring order to the chaos, nobody could have rallied such broken troops. Soon, heavy fighting had broken out on both sides of the bridge. Everywhere the cry went up: "The Turks are here; all is lost; save yourselves." Meanwhile, junior officers were actually bringing their units into action against the unseen foe. Volley fire broke out across the river, cannons fired into the inky void, while swarms of cavalry raced up and down the long columns of infantry, hacking and slashing at imaginary Turks. By the time order was restored with the morning light, the Austrian army—victim of mass hysteria—had suffered thousands of self-inflicted casualties.

THE NINETEENTH CENTURY

During the Napoleonic Wars, with the enormous increase in the size of armies and the consequently larger battles—the Battle of Leipzig in 1813 was not to be exceeded in numbers engaged until the First World War—there were inevitably many examples of accidents leading to friendly casualties in the armies of all the combatant nations. Armies were still the products of eighteenth-century drill practices, and the problems of that system still lay heavily on the armies of England, Austria, Russia, and Prussia, though the French were introducing more flexible methods.

Carelessness was at the root of numerous accidents causing friendly casualties or self-inflicted wounds. Peninsular War veteran John Green describes one sort of mishap that could occur as easily in battle as on the parade ground:

> About three weeks before the half-yearly inspection, we began to prepare for it by going through our evolutions and maneuvers in the large barrack-yard: towards the latter end of the time, in order that the regiment might learn to be steady, we fired with blank cartridge. A man called Malfrey, about five men from myself to the right, of Captain Gough's company, had loaded his piece five times, it missing every time. The sergeant in the rear told him he dare not fire it off; the man declared if it was full of devils he would; he did so in the next volley and the consequences were dreadful; for his musket burst into several pieces, carrying away part of his hand, and wounding or burning several men who were near him, so that this part of our line was thrown into confusion. I saw a dog run away with one of his fingers.[6]

Captain Alexander Gordon of the Fifteenth Hussars recorded another curious accident. Engaging the French near Sahagun, in Spain, just as night was falling, Gordon heard an explosion behind him and turned back in time to witness one of his own troopers tumbling from his horse. Everyone around burst into laughter, in spite of the seriousness of the situation, because the trooper in aiming at a French dragoon had managed to shoot his own horse by mistake. Later, in the growing dusk, Gordon had less reason for merriment when he was attacked and almost killed by members of his own troop who recognized him only at the last moment before their swords descended upon his head. Careless practice by the gunners was usually fatal, though not in this bizarre example related by Private Alexander Alexander in 1811:

One occurrence I witnessed here was almost incredible: a Portuguese governor touched at Colombo early in the year 1811; on the firing of the salute, Gunner Richard Clark was blown from the mouth of his gun right into the air, and alighted upon a rock at a considerable distance in the harbor, yet escaped without a bone being broken, almost unhurt. It was the most miraculous escape I ever witnessed; he was but an awkward soldier at the best; the gun of which he was No. 1, went off by accident, but not just at the time of loading, otherwise the left arm, or perhaps both arms, of No. 2 had been blown off, as No. 2 loads and rams home, along with No. 1. The gun was just loaded when she went off, through the negligence of Clark, in not spunging properly. He was not at his proper distance, like the other man, nor yet near enough to receive the whole flash. To the astonishment of everyone, he was seen in the air, the spunge-staff grasped in his right hand, the rammerhead downwards, which first struck the rock as he alighted on his breech. The rock was very thickly covered with sea weed. A party was sent down to bring up the body, as all concluded him killed upon the spot; he was brought up only stunned and slightly singed, and was at his duty again in a few days . . . [7]

Captain Cavalié Mercer, whose experience of friendly fire with the Prussians is told below, experienced a tragic accident at the Battle of Waterloo caused by the carelessness of one of his gunners. The man stumbled after loading his gun at the very moment of firing: "As a man naturally does when falling, he threw out both his arms before him, and they were blown off at the elbows." The man later bled to death before he could get medical attention.

If the British had their problems with training, the recruits of Napoleon's Grande Armee were not exempt, particularly in the early stages of their military career. In 1804, at Leghorn, a French soldier forgot to remove his ramrod during practice firing. When he fired his musket, the ramrod shot out like an arrow, killing a spectator who, ironically, turned out to be a criminal wanted by the local police. Thus the incompetent recruit was rewarded rather than punished for his extraordinary example of friendly fire. An even worse incident of the same kind occurred at Valladolid, in Spain, four years later. Again blood was spilled on the parade ground while General Jean

Malher was drilling a unit of raw recruits. He made the cardinal mistake of walking in front of the men when they fired their blanks. Unfortunately for Malher, no less than eighteen of the recruits forgot to remove their ramrods and one of them impaled him through the chest.

On the Napoleonic battlefield itself, the fog of war was all too real. Rifleman Harris in his *Recollections*, describes the limited perspectives of a soldier of the period, with all the consequent chances of accidents or friendly fire. Harris was at the Battle of Vimeiro in 1808:

> I myself was very soon hotly engaged, loading and firing away, enveloped in the smoke I created, and the cloud which hung about me from the continued fire of my comrades, that I could see nothing for a few minutes but the red flash of my own piece amongst the white vapour clinging to my very clothes. This has often seemed to me to be the greatest drawback upon our present system of fighting; for whilst in such a state, on a calm day, until some friendly breeze of wind clears the space around, a soldier knows no more of his position and what is about to happen in his front, or what has happened (even mongst his own companions) than the very dead lying around . . . Often I was obliged to stop firing . . . and try in vain to get a sight of what was going on.[8]

In the smoky chaos that Harris describes, accidental wounding of comrades and allies must have been widespread. In the great Napoleonic setpieces, notably those like Waterloo where infantry squares were formed to repel cavalry attacks, it was inevitable that front-rank soldiers would be raked by stray bullets from neighboring units. Richard Holmes has cited an example from the Franco-Prussian War in 1870, where, at the Battle of Rezonville, Lieutenant Devauriex of the Sixty-sixth Regiment commented that his men were firing as if "they were drunk on rifle fire during the gripping crisis." His men were eventually reproached by the brigadier of a neighboring regiment who gently called out to them, "Mes enfants, you probably do not realize it, but you are firing on my brigade."

At the Battle of Talavera in 1809, during the Peninsular War, the Spanish forces of General Cuesta, allied to the Sir Arthur Wellesley's British army, were facing the French. Although the enemy were obviously well out of range, Cuesta's infantry insisted on firing a tremendous volley at them and then, apparently frightened by their own fire, panicked and fled. Before quitting the field, however, they took the opportunity to plunder the British camp, wiping out the various camp followers who were at that stage in possession of the wagons. One feels that to use the term "amicide" in such a situation is to stretch the definition beyond reason. It is as well to add that the seriously depleted British, now outnumbered two to one after the Spanish flight, still managed to soundly defeat the French.

Friendly casualties formed a high proportion of casualties in Napoleon's battles, often as a result of the close nature of the infantry engagements. On the late evening of the first day (February 7, 1807) of the Battle of Eylau, the fighting between the French and Russians became so confused that both sides were guilty of firing on their own colleagues. The struggle at Eylau began as an encounter battle, with neither side intending at the outset to commit their full strength. Apparently it really began when Napoleon's baggage and his personal attendants arrived in Eylau by mistake, having not been informed that the emperor was intending to spend the night some distance away at Ziegelhof. As the servants began unpacking what David Chandler refers to as the "Imperial comforts," they were attacked by a Russian patrol and would have been captured or worse had not the detachment of the guards that always accompanied Napoleon's effects charged to their aid. At the sound of firing, both sides rushed up reinforcements. Marshal Soult in person descended on the Russians, who were already trying to pillage the baggage wagons, while the Russian generals, assuming that the French were occupying Eylau in force, brought up more troops. Soon a fierce struggle was taking place in the streets of the town, and notably in the cemetery. Even after darkness fell, the fighting continued, with many friendly casualties. Each side lost as many as four thousand men in this part of the engagement, until the Russian General Bennigsen called his men back out of the town to occupy a nearby ridge. Further fighting flared up occasionally, and throughout the night mistakes were so frequent that both sides had the greatest difficulty stopping their own men firing on each other. During the dreadful fighting in the blizzards the following day, there were many further incidents of friendly fire, the most significant of which occurred when Marshal Augureau's division was hit by a barrage fired by the French artillery, which had been blinded by the snow.

During the great Battle of Wagram in 1809, French troops opened fire on their Saxon allies, whose gray uniforms were misidentified by the French as the white coats of their Austrian enemies. In the opening stages of the Battle of Busaco in Spain in 1810, Wellington's relatively inexperienced Portuguese troops—notably the Eighth Infantry Regiment and a battalion of militia—fired several volleys by mistake into the ranks of their British allies. In fact, the Peninsular War was the scene of many such friendly fire incidents. At the Battle of Albuera the following year, the British Twenty-ninth Foot inflicted heavy casualties on their Spanish allies. Having marched to the rescue of General Beresford and his staff, the Twenty-ninth then opened a heavy fire on the dispersed French lancers. Unfortunately, their fire was wayward, and instead of hitting the French cavalry, they shattered the Spanish troops of General Zayas instead. During the bloody siege of Badajoz in 1812, there was so much confusion that the British redcoats fought a lengthy firefight among themselves. On the St. Vicente bastion, members of the Fifth Division first cleared away French resistance, only to find themselves under fire from the British Third Division, which had entered the city from the other side. In the darkness, men of General Walker's brigade kept up an exchange of shots with General Picton's redcoats, until dawn revealed the true identity of their assailants.

In the opinion of the French marshal, Gouvion St. Cyr, a quarter of all French soldiers who died during the Napoleonic period, were killed in battle by their own artillery and musket fire rather than by those of the enemy. Exaggerated though this view may be, it is interesting to hear it repeated on the field of Waterloo by the colonel of the British Twenty-third Light Dragoons: "It's always the case, we always lose more men by our own people than we do by the enemy." We can excuse the colonel's frustration—he had just had his horse killed under him by one of the British redcoats. Nevertheless, a study of the fighting at Waterloo on June 18, 1815, enables us to tell just how widespread was the problem of friendly fire. Through the efforts of Captain William Siborne, who circularized all the British officers who had survived the battle, we are able to penetrate the fog of war and construct a view of the battle as seen from the British side. The picture thus created is one of considerable confusion and numerous accidents.

The arrival of the advanced Prussian corps in the late afternoon of June 18 posed almost as much of a problem to the Duke of Wellington as to Napoleon himself. Both the Prussian artillery and the cavalry immediately joined battle with sections of Wellington's command. Captain Mercer, commanding a British horse artillery battery, recounts in his journal one of the most famous—and most amusing—examples of friendly fire ever recorded.

Mercer and his men had just helped to repulse the massive French cavalry charges on the British infantry squares ordered by Marshal Ney and were, no doubt, reflecting on a job well-done, when suddenly they came under heavy fire from a most unexpected quarter. A Prussian artillery battery, wrongly identifying Mercer's guns as French, had opened fire without warning and was soon inflicting heavy damage on the British guns. As Mercer wryly commented: "The whole livelong day had cost us nothing like this." Exhausted as were all of the British gunners by the titanic struggle against the French, it would have been a perverse chance to succumb to one's supposed allies rather than the enemy, and Mercer soon decided to fight back, turning his guns to fire at the new "enemy." Hardly had he done so than a rider, dressed in the black uniform of Brunswick, rode up to his battery calling at the top of his voice: "Ah! mine Gott!—mine Gott! vot is it you doos, sare? Dat is your friends de Proosiens; an you kills dem! Ah mine Gott!—mine Gott! vill you no stop, sare?—vill you no stop? Vat for is dis? De Inglish kills dere friends de Proosiens! Vere is de Dook von Vellington?" Mercer, with appropriate British understatement, observed that the officer was "raving like one demented." He then suggested to the Brunswicker "that if these were our friends the Prussians they were treating us very uncivilly; and that it was not without sufficient provocation we had turned our guns on them." Mercer agreed to stop firing for a while to see if the Prussians would recognize their error and also stop firing. However, with the German officer standing beside him, Mercer was forced to dive for cover as more Prussian shells arrived. He now prevailed on the German in no uncertain terms to ride back and tell his friends that the British would stop firing when they did—and not before. The bemused German returned to his horse, muttering "Oh, dis is terreeble to see de Proosien and de Inglish kill vonanoder!" Mercer relates that his problem was solved for him not by the German officer's mediation but by a further bizarre intervention. A battery of guns manned by Belgian troops whom Mercer describes as "beastly drunk and . . . not at all particular as to which way they fired" mistook the Prussian gunners who were firing at Mercer for Frenchmen and promptly shattered them with shot and canister. As Mercer reflected: "The Belgians . . . would have fired on us too, had we not taken pains to put them straight."

While the above farce was taking place, on another part of the field an even more costly friendly engagement was occurring. As the first Prussian troops came up in support of Wellington's left flank, they encountered Nassauers, commanded by the Duke of Saxe-Weimar, wearing the same uniforms that they had worn only twelve months before when they had fought the Prussians as allies of Napoleon. To compound the mistake, the Duke of Saxe-Weimar, assuming that the newcomers were men from the French Marshal Grouchy's detached corps, opened fire on them, and a savage fight ensued. The Nassauers who had held their defensive positions all day, were now driven from the line by the Prussians, with heavy losses on both sides. Even though the tragic error was eventually pointed out, the Nassau troops were too exhausted to return to their positions and took no further part in the battle.

While the Prussians were engaged in fighting both friend and foe alike, other revealing but otherwise less significant "encounters" were taking place in other parts of the field. Siborne's research has thrown up a number of incidents that must have been replicated on battlefields throughout the eighteenth and nineteenth centuries, wherever troops fought in the open armed with muskets, bayonets, swords, and lances. In the congested battlefield of Waterloo—one of the smallest in area of all Napoleonic battles, consequently producing one of the bloodiest of all such engagements—thousands of men on both sides must have fallen to friendly fire. Only discipline, in the shape of a mind-numbing regime of drill and more drill, kept soldiers of this period from cutting swathes through their own ranks before ever coming to grips with the enemy. And in such a hard-fought battle as Waterloo and with so many different nationalities present in Wellington's army—British, Portuguese, Belgian, Dutch, and men from Hanover, Brunswick, and other principalities—it is hardly surprising that discipline faltered and accidents occurred.

With the help of Siborne, we know that in the last stages of the battle, with discipline stretched beyond its limits and the French army disintegrating, many accidental assaults took place in the growing darkness. Colonel Hay of the Sixteenth Light Dragoons was shot by British infantry, while the Tenth Hussars were fired on by British artillery, who had been firing at the same French cavalry that the Hussars were pursuing. The Twelfth Light Dragoons clashed briefly—fortunately at the cost of only a single injury—with the First Hussars of the King's German Legion. In the confusion of a stricken field, the Sixteenth Light Dragoons only narrowly averted clashing with the same body of German Hussars in the growing darkness. A little earlier—though still late in the day—the Eighteenth Hussars of Vivian's brigade clashed with Prussian cavalry near La Belle Alliance, and some of the latter were sadly cut down in the confusion. Nor were the infantry guiltless. As the Fifty-second Regiment—part of Adam's Infantry Brigade—was advancing at the end of the battle, it encountered two or three squadrons of the Twenty-third Dragoons and, supposing them to be French, fired into them, unhorsing Lieutenant General Lord Seaton, who came out of the incident unscathed and wrote amiably of it to Siborne in 1843. In the clashes between British and French horsemen, many accidental woundings took place. The crowding together and intermingling of enemy and friendly horsemen, combined with the nature of the weapons employed—heavy sabers or lances, both of which required space to be employed successfully—made friendly casualties inevitable. As Trooper Tomkinson of the Sixteenth Light Dragoons recorded: "Lieutenant Beckwith . . . stood still and attempted to catch this man on his sword; he missed him and nearly ran me through the body. I was following the man at a hard gallop."

Hardly a generation after the Napoleonic Wars, the problems of controlling troops in a night operation were well-illustrated in an incident during the French invasion of Algeria. The French had landed at Sidi Feruch and had established a camp on the beaches. However, at two o'clock in the morning, the sound of a loose horse running in front of the line alerted a sentry, who fired wildly into the darkness and sounded the alarm. Soon, hundreds of men were awake and rushing to get their muskets. In a matter of moments, there was a blaze of musketry all along the beach that lasted for more than fifteen minutes. In the confusion, four men were killed and ten others wounded, although there was no sign of any enemy troops. It had been a panic about nothing.

During the First Afghan War in 1839, Color Sergeant John Clarke of the Seventeenth Foot remembers an accident that occurred on the march to Kabul:

> About the fifth day's march (we were on the advance guard) our artillery came galloping up to us about daylight.
>
> We halted to let them pass, and they told us there was a battery on the plain and they were to take it. They went through a little pass to Paradise Valley. When we got to the pass we heard a loud report and thought they were engaged, but when we got through nobody was in sight but our own artillery. We hastened up to them and found a beautiful half-moon battery formed round the pass.
>
> The report we had heard was caused by one of our artillerymen, who had got on the ammunition-box and was lighting his pipe when a spark had got into some loose powder. This had exploded and blown him away.[9]

THE AMERICAN CIVIL WAR

The incidence of friendly casualties was notably high during the American Civil War. Large numbers of eager civilian recruits needed to be quickly taught the rudiments of the military profession, and this resulted in some alarming gaps in some men's knowledge. Benjamin Andrews of the Fourth Connecticut Infantry remembers how little training the average Union soldier had before facing the dreadful person of "Stonewall" Jackson and his Confederates:

> At last we had formed line, and the Colonel, on the ground probably that more battles are won by marching than by fighting, started us, raw levies, with six long miles and probably a battle before us, off on a double quick. We ran a mile, puffing, sweating, straining our eyes to see that foe we so longed to annihilate. "Halt!" What for? Why, the line officers have held a council of war while trotting along on their horses, and have concluded that if we are to fight it may be well to have our muskets loaded. No one had thought of it before. We had supposed that our brave Colonel, in whose skill as a tactician we had the most unhesitating confidence, intended on meeting Jackson, to charge with the bayonet. We concluded that he now alters his mind. At all events he commands to "load." But we have no instructions in loading. Which end of the cartridge shall go downwards? About a third of the men, reasoning apriori that the bullet was the main thing, put it in first. A good number of those who did not do this failed to tear the cartridge paper. Several put two or three cartridges in; some even more. It was the work of a week to empty these muskets. Having loaded and breathed we began to race again.[10]

Even as late as the Battle of Gettysburg in 1863, faulty loading was a feature of the fire discipline. After the fighting had ended, Union troops tried to clear the battlefield, finding there an incredible 27,574 abandoned muskets and rifles, of which twenty-four thousand were loaded. Of these, twelve thousand were loaded twice, six thousand were loaded between three and ten times, and one boasted twenty-three charges and no balls, little less than a bomb rather than a firearm. Every kind of curious combination of ball and powder charge had been attempted, including one musket loaded with twenty-two balls and sixty-six buckshot. Many had the ball behind the charge and could not have fired whatever the soldier did. Paddy Griffith in his book *Rally Once Again* has concluded that some 9 percent of all muskets were misloaded. According to Griffith: "A very high proportion of infantry weapons must indeed have become inoperative in combat due to faulty handling." And a consequence of this was that many friendly casualties were suffered when these muskets exploded or misfired, as so many of them obviously did in the heat of battle.

> The real problem with Civil War infantry fire always lay in the most rudimentary aspects of maintaining the men's concentration on their job. In combat, fire would usually be delivered from a line two ranks deep, with the soldiers in each rank almost touching each other's elbows. They would naturally jostle and shove each other as they drew their ramrods and pushed home their cartridges. Anyone in the second rank would have to lean forward to fire through the space between two men in the front rank, who would receive a flash and a cloud of smoke in their eyes and a numbing explosion at the level of their ears.[11]

At the First Battle of Bull Run (Manassas) in 1861, the new recruits on both sides found it immensely difficult to maintain tight discipline. One Union soldier remarked: "The men were a good deal excited. Our rear rank had singed the hair of the front rank, who were more afraid of them than of the Rebels." A Confederate soldier sees things in very much the same way:

> A battle is entered into mostly in as good order and with as close a drill front as the nature of the ground will permit, but at the first "pop! pop!" of the rifles there comes a sudden loosening of the ranks, a freeing of selves from the impediment of contact, and every man goes to fighting on his own hook; firing as, and when he likes, and reloading as fast as he fires . . . A battle is too busy a time, and too absorbing, to admit of a good deal of talk. Still you will hear such remarks as "Looky here, Butler, mind how you shoot; that ball didn't miss my head two inches" . . . [12]

General Sherman noted that during the Civil War, perfect fire discipline in battle was hardly ever attainable:

> Very few of the battles in which I have participated were fought as described in European text-books, viz., in great masses, in

perfect order, maneuvring [sic] by corps, divisions, and brigades. We were generally in wooded country, and, though our lines were deployed according to tactics, the men generally fought.[13]

During the Battle of Seven Pines in 1862, the woodland setting contributed to much confusion on both sides. Units would appear from thick foliage and encounter "enemies" at close range. This being the case, it was rarely possible to make an early identification of the opposition, and it was often impossible to separate two friendly units before blood had been spilled.

The Twenty-fourth Michigan, formed in July 1862, was rushed into action before it had received proper training. On the only occasion it was allowed a target practice, three men were wounded in "accidents" and one died of a heart attack. In Paddy Griffith's opinion: "Live fire was almost as dangerous to the men who were delivering it as it was to the enemy." If this standard of fire discipline is to be considered as typical of the hastily raised Union forces, it is not difficult to imagine why friendly fire casualties were so heavy in the often confused fighting that typified the Civil War period. As Griffith has demonstrated: "An almost total lack of target practice meant than many rifles were misloaded in combat and that the finer points of long-range accuracy were neglected or ignored. The close-order drill of the day also meant that the soldier in battle was subjected to a barrage of sights, sounds and emotions which must have distracted him powerfully from his task." It was not possible—either in the United States or in Europe—to transform an essentially civilian population into a military one without a prolonged period of training. The United States, like Great Britain, had a fear of standing armies and preferred to raise its military forces when the need arose. Thus the number of well-trained regulars available in the United States or in Britain was tiny when compared with the conscripted mass armies available in France and Germany. The transformation of American civilians in 1861–62 into soldiers resembles the creation of Britain's first civilian army by Lord Kitchener in 1914–15. The raw material was good, but the process of metamorphosis was often a long one, and was not achieved without numerous hiccups, in terms of friendly fire and camp "accidents."

In John Pullen's *The Twentieth Maine,* several of these camp accidents are described even late in the war, when one might have assumed that the regiment had overcome such teething troubles. During the winter of 1864, Pullen describes a lowering of discipline through inactivity. Scuffles were taking place over slight or imagined insults, drunkenness was more common, and there was a general air of restlessness. In this atmosphere, soldiers become careless. One private, we are told, was shot through the head by his tentmate, who aimed what he believed to be an empty rifle at him and pulled the trigger. In the last few days of the war, a careless wagoner accidentally "discharged a carbine." The bullet apparently passed through several tents before killing Lieutenant George Wood, a brave officer who had fought with the regiment all the way through the war.

At First Bull Run, Confederate Brigadier General John Imboden nearly paid with his life for his carelessness. As he wrote:

> Lieutenant Harman and I had amused ourselves training one of the guns on a heavy column of the enemy, who were advancing towards us, in the direction of the Chinn house, but were still 1200 to 1500 yards away. While we were thus engaged, General Jackson rode up and said that three or four batteries were approaching rapidly, And that we might soon retire. I asked permission to fire the three rounds of shrapnel left to us, and he said, "Go ahead." I picked up a charge (the fuse was cut and ready) and rammed it home myself, remarking to Harman, "Tom, put in the primer and pull her off." I forgot to step back far enough from the muzzle, and, as I wanted to see the shell strike, I squatted to be under the smoke, and gave the word "Fire." Heavens! what a report. Finding myself full twenty feet away, I thought the gun had burst. But it was only the pent-up gas, that, escaping sideways as the shot cleared the muzzle, had struck my side and head with great violence. I recovered in time to see the shell explode in the enemy's ranks. The blood gushed out of my left ear, and from that day to this it has been totally deaf.[14]

The unintentional killing of officers, as opposed to deliberate "fragging" (see p. 232), has always been a feature of large-scale military engagements. With the advent of musketry and long-range cannons, the commander, whatever protection he might have from bodyguards or aides, became as vulnerable as any other man on the field. By his uniform or by the elevated position he occupied to get a better view of the fighting, he might even be more noticeable to the enemy. Yet it has been in the metaphorical fog of war that commanders have succumbed to the fire of their own, sometimes disoriented, troopers. During the American Civil War, several prominent officers were shot by their own side, by far the most famous of whom—an irreplaceable loss to the Confederacy—was General Thomas "Stonewall" Jackson, the most renowned of Robert E. Lee's lieutenants. Lee always claimed that it was Jackson—his "strong right arm"—who contributed most to his greatest victories. Certainly with the death of Jackson, Lee seemed to lose his capacity to virtually hypnotize Union commanders. In 1863, at the Battle of Chancellorsville, Jackson had literally run rings around the Union commander "Fighting Joe" Hooker and had laid the foundation for Lee's most perfect victory. Yet for Lee, Chancellorsville was a Pyrrhic victory when news reached him that Jackson had been mortally wounded by his own men.

One of Jackson's aides-de-camp, the Reverend James Power Smith, was with him when he suffered his fatal accident and wrote a full account of what happened. About a mile to the west of Chancellorsville, Jackson found his front-line troops little more than half a mile from the Federals. It was dusk, and in the poor light, made even worse by the dense thickets of undergrowth, Jackson was finding it difficult to correctly align the troops of A. P. Hill's division. With just two or three of his staff and a small group of couriers and signalers, Jackson rode down the turnpike in the direction of the Federal lines before encountering their pickets and turning back toward his own lines. It was this turn backward that was to prove fatal. In the growing darkness, all the Confederate troops could identify was a body of horsemen riding toward them from the direction of the Federal lines. Some of the Confederate troopers opened a ragged fire at Jackson and his band, and two of the riders, an engineer officer named Captain Boswell and a signaler named Sergeant Cunliffe, were shot dead as they approached. Jackson and the survivors veered away, only to be hit by a second volley of fire from a company of General Pender's North Carolina Brigade. Jackson was hit by three separate balls, one through the palm of his left hand, another through his left wrist, and the third through his left upper arm, splintering the bone from shoulder to elbow. His horse turned away from the fire and rushed off, dragging the wounded general into thick bushes.

Before Jackson could fall from his horse, he was caught by a signals officer, Captain Wilbourn, and gently lowered to the ground. Almost at once, General A. P. Hill rode up with his staff and dismounted at his prone leader's side. The Reverend Smith staunched the flow of blood from the wound in the upper arm, but it was very difficult to move Jackson from such a forward position to an ambulance. Federal guns were sweeping the whole area with canister shell, and it was impossible for the litter bearers to carry the general out of danger. Even in his desperate state Jackson did not forget his responsibilities. When General Pender remarked that he would have to withdraw his troops to re-form them, the injured commander replied: "You must hold your ground, General Pender; you must hold your ground, sir." As the Reverend Smith observed, it was his last command on a battlefield and it was in keeping with his nickname of "Stonewall." Once Jackson reached an ambulance, he was taken back to a field hospital where his good friend, Dr. Hunter McGuire, amputated his left arm just below the shoulder. At first, it appeared that the general would make a complete recovery, and a dispatch was sent to the Confederate commander in chief, Robert E. Lee, informing him that his "strong right arm" was wounded. Lee apparently received the news "with profound grief" and replied: "Could I have directed events, I should have chosen, for the good of the country, to have been disabled in your stead. I congratulate you upon the victory which is due to your skill and energy." Typically, Jackson preferred to attribute the victory to God's work rather than man's.

At first, the doctors were hopeful of Jackson making a complete recovery, but in spite of every effort, pneumonia set in, and on May 10, 1863, the great Stonewall Jackson died. His last words apparently were: "Let us pass over the river, and rest under the shade of the trees . . ."

The death of Jackson was a blow to the whole Confederacy. Had he been at Lee's side at the Battle of Gettysburg, a Confederate victory would have been probable on the first or second day of fighting. But for Jackson's untimely end, the whole history of the United States might well have taken a different course. Nor was Jackson the only Confederate general to suffer at the hands of trigger-happy Confederate troops. During the Wilderness fighting, on May 6, 1864, "Old Pete" Longstreet—another of Lee's most able lieutenants—was severely wounded by friendly fire. Longstreet was riding with Brigadier General Micah Jenkins at the head of his brigade when they encountered units from Major General William Mahone's division. Because they failed to identify each other as friends, a ragged exchange of fire took place between Jenkins's men and Mahone's, during which Brigadier General Jenkins was killed and General Longstreet was seriously wounded.

The Twentieth Maine was also plagued by that bane of all modern infantry—short firing by friendly artillery. At the Battle of Ball's Bluff, one of the soldiers, Private William Livermore, wrote: "The shells from our batteries would go so near our heads it seemed as though it would take the hair off from my head, and the air was full of shells and some of our own burst overhead and wounded some of our own men." But Livermore seems to have taken this friendly fire in his stride and expresses no criticism of the Union gunners. Yet at this stage many of the artillerymen deserved more than just verbal criticism. A few days later, the Twentieth Maine was again the victim of short firing. During a bombardment of Confederate positions, one artillery battery had set its fuses so badly that its shells were bursting after only 50 percent of their trajectory. The result was that they were bursting as they passed over the men from Maine, and many of them were wounded by a shower of metal. Eventually, a delegation was sent back to reason with these gunners and, in the words of John Pullen, the guns "fell silent with dramatic suddenness."

THE SPANISH-AMERICAN WAR

The problem of undisciplined artillery fire was a feature of the ground fighting during the war between the United States and Spain in 1898. The high number of friendly fire incidents that occurred was a reflection of the chaotic nature of the conflict and the poor training of the troops on both sides. The main land campaign took place in Cuba between the numerous but poorly motivated Spanish troops and the exuberant but undisciplined Americans. As a result, it was remarkable for some of the grossest inefficiency ever witnessed in the history of the American army.

The Spanish capital on Cuba, Santiago, was protected by strong Spanish entrenchments on San Juan Hill, which the approaching U.S. troops would need to assault before they could advance on the city. On July 1, two divisions under Generals Sumner and Kent moved through thick jungle toward the base of San Juan Hill. On a hill near El Pozo, Captain Grimes, commanding the U.S. artillery, established his batteries to support the infantry assault. As he did so, he attracted a large crowd of tourists and onlookers. Grimes, an incompetent gunner to say the least, was still using black powder, and his guns produced so much smoke that the Spanish counter batteries were able to locate him easily and force him to stop firing. As Walter Millis wrote:

> Congress had neglected to provide our artillery with the modern smokeless powder, and as the first great clouds of white smoke billowed forth from El Pozo, the Spaniards very naturally took them as a target for their own artillery. The cameras recording the first "shot" were still clicking and an interested crowd of people from the regiments below was just gathering upon the hill to see what was going on, when the first answering shell sang over the battery and burst on the slope behind it, extinguishing a number of Cubans and wounding several Rough Riders who were in the farmyard below.[15]

For a while, the American infantry advanced peacefully through the jungle, but when they reached the bottom of San Juan Hill, they faced a daunting prospect of having to attack a heavily entrenched enemy without artillery support. Fortunately, a battery of Gatling guns commanded by Lieutenant Parker opened fire on the Spanish positions, causing many of the defenders to fling down their weapons and take to their heels. Seeing this, the American infantry now stormed up the slopes of San Juan Hill. The scene was set for Captain Grimes to make a serious contribution to the American victory. From El Pozo, he had been unable to follow the fighting very well and so was unable to identify the figures who began to appear on the green slopes of the hillside as blue dots. Wrongly assuming them to be Spaniards, he decided to join the battle. His guns were trained on the slopes of San Juan Hill and opened fire with farcical consequences. As Captain Allen of the Sixteenth Infantry wrote:

> The advance continued steadily and without a pause until we were on the steep slope near the crest, two thirds of the way up, when our artillery fire coming from our rear became dangerous . . . Some shells struck the slope between me and the crest . . . there arose at the foot of the slope and in the field behind us a great cry of "Come back! Come back!" The trumpets there sounded "Cease-firing," "Recall" and "Assembly." The men hesitated, stopped, and began drifting down the steep slope . . . [16]

One officer had the initiative to wave his hat at the gunners, whereupon they fired and wounded him. It took some time before the artillery could be silenced so that the advance could continue. What would have happened had the Spaniards counterattacked at this moment and recaptured the hill is perhaps futile to ask. But Captain Grimes and his guns had contributed not only to friendly casualties—light as these were—but to repulsing a major attack by his own troops on a vital enemy target.

2

GROUND WARFARE FROM 1914 TO THE PRESENT DAY

THE FIRST WORLD WAR

In the history of modern warfare, there can be little doubt that artillery has made the most devastating contribution to friendly fire casualties. In the First World War, the gunners of all armies were hated by the frontline troops who suffered so terribly from the "shorts" that fell regularly into their trenches from their own guns. Relations were sometimes so bad between the gunners and the "poor, bloody infantry" that fights broke out behind the lines and in estaminets in rear areas. The attitude of the infantry can be well-understood. Their fate was hard enough, facing the shells and machine guns of a vigilant enemy, without having to fear for their backs. Yet the gunners were themselves victims of a new kind of warfare, both in scale and technique, and the degree of precision required for close cooperation with ground troops was beyond anything that had ever been asked of artillerists before. So if there were friendly casualties aplenty through human error, they were often a product of technology moving beyond the capacity of the human beings to control it.

The demands on artillery in the twentieth century have indeed been far greater than ever before. Gunners, who in Napoleonic times were as much battlefield troops as the infantry and took their chance among the bayonets and sabers of the enemy like anyone else, were pushed far back from the front lines by the increasing ranges of their own guns and became indirect rather than direct participants on the battlefield. The infantry never saw their struggles, thousands of yards or even several miles behind the front lines, against an unseen enemy who sought their lives as eagerly as those of the men in the trenches. And fighting an unseen enemy was always fraught with difficulties. Indirect rather than direct fire was bound to increase the number of mistakes by which friendly forces might be engaged through errors of map reading or even as a result of a rapidly changing tactical situation. The advent of aircraft that could report back the fall of shell to the gunners was an advance but was still too slow to guarantee that targeting of the guns was efficient. A published report of the performance of the French artillery offered a sharp warning against complacency. Statistics for British friendly fire casualties simply do not exist, but there are numerous examples of artillery amicide by British gunners recorded in diaries and military memoirs, some so dreadful that their details have been expunged as far as possible from any official records.

In 1921, an artillery expert and critic of the French high command, General Percin, published the first and most comprehensive account of friendly fire in military literature. He entitled his book *Le Massacre de notre Infanterie, 1914–1918* and in it claimed that as many as seventy-five thousand French casualties had been caused in friendly fire incidents involving the French artillery. It was an astonishing claim, yet he was supported by evidence from hundreds of correspondents, reporting from all periods of the war and from all parts of the French lines on the Western Front. There seems little doubt that a British or a German book of like kind could have been produced had an Anglo-Saxon or Teutonic Percin set his mind to it. That they did not is our loss, yet as I attempt to show below, numerous examples of friendly fire on the Anglo-German front on the Somme and in Flanders can be assembled by a close study of the huge literature of military memoirs and personal accounts by officers and men alike. Understandably, perhaps, the official histories—with the exception of C. E. W. Bean's Australian history—are less forthcoming about friendly casualties. When they were written, the idea of heavy losses caused by carelessness, stupidity, and incompetence was still far from an acceptable one. Nevertheless, it would not be stretching the bounds of credulity to assert that the static warfare in France and Belgium from 1915 to 1918 and the preeminent role of the big guns provided opportunities for mistakes on an unparalleled scale. The degree of precision required of both the guns and their gunners was frequently missing before technological and professional skills reached their height in 1918. Thus the massive British artillery bombardments on the Somme in June-July 1916 and before the Third Battle of Ypres in 1917 produced both a tragic and yet understandable quota of friendly fire incidents.

There is a sameness about many of the French incidents Percin relates that must infuriate the sensitive reader, aware that the cost was being paid in human lives and suffering, because so little seems to change between his first example on August 15, 1914, and his last, on November 1, 1918, just ten days before the armistice. It seems that nothing was learned from the massacre of the French infantry and that the kind of errors responsible for decimating the exuberant poilus in their *pantalons rouges* in the first few days of the war were those that scythed down the grim-faced, cynical, and mutinous veterans fifty months later.

On August 22, 1914, Percin tells us, the First Regiment of Colonial Infantry, led by Colonel Guérin, was moving toward the town of Neufchateau in Belgium. Their path took them through dense forest near Rossignol. Information from the local people told Guérin that there were German troops in the forest, yet, undeterred, the colonel marched in only to find himself confronted by three regiments of German soldiers, well dug-in. A bloody and confused fight took place that raged for hours. A battery of artillery that had accompanied Guérin's regiment set up its guns four hundred meters from the forest and, lacking orders of any kind, fired blindly into the trees in apparent support of its infantry. But the gunners had no way of knowing where the French troops were at any one time, and the battle became a lottery. By the end of the day, out of 3,250 poilus who had entered the forest, 2,000 were dead and a further 1,000 wounded or taken prisoner by the Germans. In the words of one of the survivors: "Our artillery went mad." It is estimated that more than a third of all the French casualties were caused by their own guns, including nearly seven hundred men killed. It had been a shocking example of the breakdown in liaison between different elements of an army. How had it been possible for the gunners to continue firing when they had no orders from a senior officer either to open fire in the first place or to continue firing without precise directions? This question remains unanswered both in this case and in the majority of examples given in Percin's book.

On December 20, 1914, French 75mm guns, situated five hundred meters behind the front lines, shattered an attack by the Thirty-eighth Regiment of Colonial Infantry in the appropriately named district of Calvary near Beausejour, on the Marne. Throughout the day, the poilus, reinforced later by the Thirty-third Regiment, clung to a toehold in the German lines, while they were alternately bombarded by the German heavy guns and raked by their own artillery. It is estimated that 35 percent of their casualties were a direct result of friendly fire by their own gunners, who showed no thought for their plight but continued firing throughout the day, in Percin's words, "without orders, without signals and without an officer present."

Short shelling by the French gunners, caused by damp ammunition, worn gun barrels, unfavorable wind conditions, and any number of other reasons—even including bad blood between rival officers of regiments—was not something the French kept entirely to themselves. Occasionally, the French gunners fired on British troops holding adjoining positions. Jean Giraudoux once boasted to Paul Morand: "I belong to the French regiment that has killed the most English." It would be surprising if some British gunner was not able to boast as much of the French he had killed.

For France, the tragedy at the village of Samogneux has come to symbolize the true horror of friendly fire. On February 21, 1916, the Germans began their assault on the French fortress of Verdun. At Samogneux, the 351st Regiment of the Seventy-second division, commanded by Lieutenant Colonel Bernard, heroically held up the German advance for two days. Hemmed in on all sides by Germans, Colonel Bernard found it increasingly difficult to communicate with his headquarters. As he wrote in the last message he managed to get through:

"All the horses have been killed, bicycles smashed, runners wounded or scattered along the routes. I shall be doing the impossible if I keep you informed of events." After this, divisional headquarters received no further word from Bernard. The only evidence of what was happening at Samogneux was gleaned by Major Becker from the words of a courier, who rode past him at full gallop shouting: "The Boche is at Samogneux." Becker was unable to question the man further and so assumed that the worst had happened and that Bernard's resistance was ended. It was, after all, very likely that the Germans had in fact taken Samogneux. When this news was passed to General Herr in Verdun, he ordered the village to be recaptured immediately. Before an assault could be launched, however, it would be necessary for the French artillery—155mm guns—to saturate the area and destroy the advanced German positions. On the night of February 23, a massive artillery barrage rained down on Samogneux. Unfortunately, at the very moment that the first French shells began to land, Colonel Bernard had at last found a way of sending a message to tell HQ that he was still holding on. It was too late. In spite of the fact that Bernard's men fired green cease-fire rockets to try to stop their gunners, it was to no avail. For once the French gunners showed unerring accuracy and the French defenders were massacred. Unknown to the gunners, they were making matters easy for the Germans, who were able to walk into the village unchallenged. The sight that met their eyes was astonishing. The entire French garrison had been wiped out by their own artillery, except for one man. As they stepped over the rubble, the Germans heard a weak voice saying: "*Pour mes enfants, sauvez-moi!*" By an incredible chance, Colonel Bernard alone had survived the inferno. He was rescued and brought before the Kaiser himself. When questioned by the German emperor, he defiantly replied: "You will never enter Verdun." And he was right. Yet the defense of Samogneux had shattered the 351st Regiment of Infantry, which suffered 80 percent casualties. How many men died in the ruins of the village from General Herr's guns, we will never know. But the effect of friendly fire had not only cost France hundreds of lives, it had also presented to the Germans a vital strategic position in their efforts to capture Verdun.

In case one should suspect that the efficient German artillery never committed such errors as their French counterparts, it is as well to point out that there was much short shooting by the German guns throughout the entire war. More skilled in their creeping barrage than either the French or the British, the Germans still had one unit—the Forty-ninth Regiment of Artillery—that was dubbed the "Forty-eighth and a half" by the front-line soldiers because of its shorts. In addition, the first firing of one of the great German railway guns killed thirty German soldiers standing nearby with its blast. It was finally only possible to fire the gun by electricity from a French farmhouse situated a quarter of a mile away.

While the massive French and German armies fought the great battle of the frontiers in August 1914, things started on a more appropriate scale for the small, highly professional British Expeditionary Force. In those early days of the war, it is still hard to believe that British casualties could be regarded as individual tragedies. The men who died still had names and identities; their deaths were tragic and significant events, not submerged by the sheer enormity of the carnage being inflicted on other fronts. But such feelings did not last long. They could not survive the revelation that, rather than dying heroically, for some noble purpose, the first British soldiers of the BEF to die in France had been killed by their own side.

The Eightieth Battery of Horse Artillery, attached to the British Fifth Division, rode up toward the village of Le Cateau at nightfall, sending out riders to scout the road ahead. The local civil guard, which was composed mainly of untrained Belgian peasants, were guarding the road, and seeing the British riders approaching and hearing them speaking in a foreign language, they panicked and opened fire, spraying bullets in all directions. The British horsemen turned back and rode hard for the British lines. Unfortunately, British infantry pickets had seen them coming at the gallop and, presuming them to be hostile, opened fire down the road, killing one man outright and mortally wounding another. To the British public, it was incomprehensible: two men dead, and the Fifth Division had not even seen a German yet.

On September 27, 1914, tragedy struck the little French village of Authuille, in the Somme region of France. A farmer named Boromée Vaquette had gone out early into his fields with his herd of cows. A little later, a platoon of French soldiers, expecting to meet Germans, had emerged from nearby woods and had dimly picked out of the morning mist a gray-clad figure, hammering wood together and apparently building a barricade. It was the farmer. But the troops had been expecting to meet Germans dressed in their *feldgrau*, and so they assumed the figure was a German soldier. Before investigating, the soldiers fired a volley at the figure in gray, who threw up his arms and fell dead. The soldiers then retreated into the woods. No more "Germans" appeared, and they ventured back to look at the body; it was the farmer Boromée Vaquette. But before they could remove the body, a unit of real German troops were sighted, who advanced toward them and set up a defensive position only yards from where the farmer's body lay. All the French soldiers could do was take the news to Madame Vaquette, the farmer's wife, and offer to retrieve her husband's body if and when the German invaders were driven away from the area. And so the body lay between the lines, hidden by grass, a victim of friendly fire.

The British attack at Neuve Chapelle in March 1915 contained several examples of artillery amicide that were pointers to what the British soldiers could expect from their grossly swollen artillery arm in 1916 and 1917. In the first case, the Second Scottish Rifles were hit by shrapnel during the preliminary bombardment of the German trenches. Colonel Story remembered: "The sickening fumes of lyddite blew back into the British trenches. Great masses of earth and huge jagged pieces of metal (shell fragments) hurtled through the air. In places, the waiting troops were covered with soil and dust." The second in command of B Company, Captain Peter Kennedy, was killed by a piece of shrapnel from a British shell. Other men were wounded, but few severely. It would not have occurred to the British soldier at this period of the war that he would soon come to hate his own gunners more than he hated the enemy himself.

The Scottish Rifles advanced along with men from the Middlesex Regiment, but as they approached an area marked "Ruined House" on their maps, they came close to being shattered by the British artillery. The gunners were exceeding their own fire plan and were firing into an area that had not been designated theirs in the planning. This failure in coordination, as John Baynes points out, was at the root of many artillery blue-on-blues. Compared with the artillery disasters at Fromelles, Poziéres, Poelcapelle, and elsewhere in the years to come, this mistake seems trifling. Apparently only one officer, Major George Carter-Campbell, was wounded. But it was a mistake nonetheless, and if it were repeated on another day and on a larger scale the consequences could be tragic. One has only to think of the fate of the Sixty-sixth Division at Passchendaele (see page 87).

The Battle of Loos

On relatively few occasions in military history, an army has employed weapons that are so intrinsically unreliable that they pose almost the same threat to the side using them as to the enemy. In such cases, friendly casualties are almost inevitable. There have been few, if any, better examples of this process than the use of poison gas during the First World War. Subject to the vagaries of wind and weather, poison gas always posed a threat not only of blowing back onto the advancing formations of troops, but also of gathering in thick clouds around the enemy trenches so that even if an initial attack was successful, it was impossible for the attacking force to occupy enemy lines without falling victim to their own gas. The first gas used by British troops at the Battle of Loos in September 1915 was a case in point. Of 50,000 casualties suffered by British forces during this disastrous battle, no fewer than 2,361 were poisoned by the first release of chlorine gas from the British lines prior to the infantry assault.

The use of poison gas by the Germans at the Second Battle of Ypres in April 1915 had come as a complete surprise to the British authorities, and it was not long before the generals were pressing for a British response. The decision was therefore taken to prepare quantities of chlorine gas for use in the proposed offensive at Loos in September. However, although it was simple to produce the poison gas, it was much harder to use it effectively in a war situation. There was simply not enough time to train men in this novel form of warfare. As was later to be the case with tanks, gas technology was misused through an inability to exploit it to its full potential.

The assault on Loos was to be carried out by Sir Henry Rawlinson's Fourth Corps of General Sir Douglas Haig's First Army. However, without adequate artillery to support the attack, Rawlinson was not optimistic about his chances. As he told Haig: "I fear heavy losses and doubt if we will get through unless the gas turns up trumps which it may do, [though] we are not very good at these new improvisations." In fact, with French commander Marshal Joffre pressing the British to widen the range of the attack, there was a danger that their gas supplies would be overstretched. A month before the planned offensive, Rawlinson visited St. Omer to attend a gas demonstration. He was far from impressed. But it was already too late to persuade Haig or Joffre that the Loos operation should be reconsidered. In view of the lack of artillery and shells for the guns, the British would be entirely dependent on the effectiveness of their gas attack.

At a conference on September 6, General Haig explained to his corps commanders the role that the gas was expected to play in the attack planned for September 25. Haig blithely spoke of the gas being carried on the wind "in front of the assaulting divisions, and [creating] a panic in the German ranks, or at least [incapacitating] them for a prolonged period." If the conditions were ideal, the British attack would proceed rapidly and capture both the first and second defensive lines of the Germans. But as Robin Prior and Trevor Wilson have pointed out: "Haig's suggestion of a panic among the enemy when confronted with gas was little more than a chimera. It was known that opposing the British at Loos were seasoned troops equipped with respirators." Under the circumstances, there was no justification for expecting the Germans to panic. After all, even without gas masks the Canadians had held their lines against a German gas attack at Ypres earlier in the year. Too much was being expected of this most unreliable of weapons. Nevertheless, given ideal conditions, the German defenders might be incapacitated for just long enough to allow the British troops to get across no-man's-land and seize their trenches. Moreover, if the German machine gunners were troubled by gas or even if they suffered reduced visibility for a short period, it might be enough to allow an attack to succeed. The gas did not need to kill or maim; it was enough if it reduced the combat efficiency of the defenders for even a matter of minutes.

The main problem in employing gas as a weapon was how to project it toward the enemy. Tests were made with gas grenades and gas shells, but these were rejected in favor of gas cylinders. Grouped together in the forward trenches, the gas cylinders would be equipped with a nozzle and a pipe so that the gas could be forced under pressure toward the German lines. However, once it had left the cylinder, the gas would be entirely dependent on the prevailing wind strength and direction. The next problem for the British planners was how much gas to use. The truth was that the British had limited stocks of chlorine gas, and once Haig was persuaded by Sir John French and Marshal Joffre to widen the area under attack, the amount of gas needed to subdue the enemy exceeded Britain's capacity to produce it at that stage of the war. By the fall of 1915, it was known that German respirators could protect a soldier for a maximum of thirty minutes, after which he would succumb to the effects of the gas. The British experts therefore decided that they would need to project gas on the German lines for a minimum of forty minutes. Only at this stage would it be safe to assume that the Germans—notably the machine gunners—would be incapacitated, and as a result it would be safe to launch the British attackers into no-man's-land. But there was not enough chlorine gas available for a forty-minute operation. The reason for this deficiency was threefold. In the first place, factory production had failed to keep up with demand. Secondly, under test conditions it had been found that the gas cylinders emptied in three minutes rather than the five minutes originally estimated. Finally, now that the area to be attacked had been extended, there would be fewer gas cylinders available for each yard of front. The lack of gas supplies mirrored the shells crisis of earlier in 1915. Haig and Rawlinson had to find some way of stretching the gas to fit the new conditions, and their solution—smoke—was imaginative if hardly innovative. It was eventually decided that the smoke—to be produced in three different ways (smoke candles, smoke bombs, and phosphorus grenades)—should supplement the gas, so that the enemy would never know whether harmless smoke or poison gas was billowing toward them at any time. They would therefore be unable to remove their respirators, and by the time their full thirty-minute resistance was up, the British would be ready to give them a last ten minutes of gas. Even though the smoke would do them no lasting harm, the British surmised that the mere fact of having to wear the respirator for so long would reduce the combat efficiency of the German soldiers and make them vulnerable to an attack by fresh British troops. Furthermore, the smoke would help to conceal the progress of the infantry in crossing no-man's-land. However, smoke, like gas, was subject to the vagaries of the wind, and this was a factor quite beyond even the most careful military planning.

Dry weather had been prevalent in the middle fortnight of September, but as the countdown to the attack began, wet and misty weather began to set in. Yet even the levels of precipitation were not as vital to the success of the attack as the direction of the wind. A gas attack depended entirely on a favorable wind direction, and Haig had to face the awesome responsibility for deciding whether to release the gas at zero-hour on September 25. His meteorological adviser, Major Gold of the Royal Flying Corps, was in an equally invidious position, being responsible for advising the First Army commander whether to risk sending tens of thousands of British soldiers into the attack behind a drifting cloud of poison gas. If the wind was too strong, the gas would be dispersed too quickly to affect the German defenders and protect the attackers; if too weak, it might hover in no-man's-land and provide a danger to the advancing British troops; and if it were to change direction at the wrong time, it could wreck the entire assault and gas thousands of British soldiers. Under these circumstances, one would have expected Gold—and Haig—to approach the problem of whether to use the gas with extreme caution. In the event, Haig's own behavior must be judged as extraordinarily rash.

On September 18, under cover of darkness, the gas cylinders—called "Oojahs" by the British Tommies—began their journey to the front line, carried in the arms or on the backs of toiling soldiers. Yet, in spite of every difficulty, not one of the cylinders was broken or damaged in transit, and by September 21, five thousand Oojahs—carrying 150 tons of chlorine gas—were in place. The artillery bombardment, which was to prepare the way for the infantry assault, began in a rather desultory way, convincing the Germans that it was merely a demonstration to distract their attention from the French lines, where the real attack would take place. In fact, it was the best the British could manage at that time.

While Haig was pondering the weather on September 24, Sir John French, the British commander in chief, arrived at First Army Headquarters to discuss with him the next day's attack. Sir John brought news that the French were due to attack at 1100 hours and asked if Haig could coordinate his own attack with theirs. Haig was unsure. He was a prisoner of the weather and explained that he hoped to be able to decide on zero-hour later that evening. At 2120 hours, Haig received the final weather report from Major Gold. It was essentially favorable: "wind southerly changing to SW or West, probably increasing to 20 miles per hour." Armed with this, Haig sent out the order that the assault was to go ahead and the troops therefore moved into the forward trenches. But in the early hours of September 25, it became obvious that the wind was not picking up as Gold had suggested it would. Haig demanded another forecast and was told that "the wind would probably be stronger just after sunrise than later in the day." This was an extremely vague piece of evidence on which to base a military action. Nevertheless, by his reliance on gas Haig had worked himself into a corner. The British artillery available was quite inadequate to support a full-scale assault, and without the gas, it would probably be necessary to call off the whole operation. This would have left the French high and dry without any British support on their flanks. In the final analysis, Haig had little choice but to order the release of the gas and hope for the best. He therefore ordered the gas attack to start at 0550, with the infantry to follow forty minutes later.

But while Haig snatched a few hours' sleep, the wind dropped to little more than a breeze, and from time to time it changed direction until it was blowing from the German lines toward the British. Major General James Foulkes, who had the unenviable job of "gas adviser" to Haig, was receiving hourly reports on wind speed and direction from all parts of the British line. The evidence that was coming in suggested that the wind would definitely be unfavorable in some parts of the line. To cover this eventuality, Foulkes had given firm orders that under no circumstances were cylinders to be switched on if an unfavorable wind was blowing at zero-hour. But the wind was fitful, and no amount of reporting could guarantee what would happen at 0550. The likeliest scenario was that the wind would be generally favorable but that gas would also be blown across the British lines and in a few places back into the British trenches. Toward dawn, this was the actual situation. In many parts of the British lines, individual officers responsible for releasing the gas faced the difficult decision whether to release the gas. Poet Robert Graves has left us a description in his autobiography, *Goodbye to All That,* of the extraordinary events in his sector. The Royal Engineers officer responsible for the gas phoned his divisional headquarters twenty minutes before the gas was to be released and reported: "Dead calm. Impossible discharge accessory [code name for the gas]." To Graves's astonishment, the reply was: "Accessory to be discharged at all costs." Hearing this, one of his soldiers commented: "Such a decision seemed suicidal, and our officers were compelled to obey against all common sense."

According to Foulkes's order, the incident Graves describes should not have occurred. To understand why this "suicidal" order was given we need to find General Douglas Haig at 0500 hours, "taking the morning air" and finding that the wind had almost completely dropped. With him was his cigarette-smoking aide-de-camp, Major Alan Fletcher, who was "smoking furiously." Haig, the nonsmoker, was content to watch the young officer and note that his cigarette smoke "drifted in puffs towards the N.E." This was reassuring. Haig admitted to having had some doubts about giving the go-ahead, but apparently the cigarette smoke clinched his decision. At 0515, Haig announced that the attack would take place. He climbed a lookout tower and noted that the wind was coming gently from the Southwest. As he later wrote in his diary: "The leaves of the poplar trees gently rustled. This seemed satisfactory. But what a risk I must run of gas blowing back upon our own dense masses of troops." If Haig had watched Fletcher's cigarette smoke more intently, as his own intelligence chief, John Charteris, had been doing, he would have seen that the smoke was drifting toward the German lines but then stopping and, if anything, coming back.

At 0550, the big guns began to fire and the gas was released. In many places, it billowed toward the German lines on a good breeze, but generally it traveled no more than twenty or thirty yards and then slowed down, hanging like a curtain across no-man's-land. In some parts of the line, individual officers had made their own decision, based on General Foulkes's order not to release the gas if there was a danger of it blowing back. Lieutenant White, Royal Engineers, of the Second Division, telephoned brigade headquarters to say that he would not open the gas cylinders as conditions were not right. He was surprised to receive an outright order from his brigadier general to release the gas. Although the general was no happier than White about the situation, he himself had received orders from higher up to carry on. White later wrote:

> At first the gas drifted slowly towards the German lines (it was plainly visible owing to the rain) but at one or two bends of the trench the gas drifted into it. In these cases I had it turned off at once. At about 6.20 a.m. the wind changed and quantities of the gas came back over our own parapet, so I ordered all gas to be turned off and only smoke candles to be used.
>
> Punctually at 6.30 a.m. one company of the King's advanced to the attack wearing smoke helmets. But there was a certain amount of confusion in the front trench owing to the presence of large quantities of gas . . . Nearly all my men suffered from the gas and four had to go to hospital. Three out of five machine-guns on my front were put out of action by the gas.[17]

The British troops, encumbered by their own gas masks, were suffering from the burden as heavily as the Germans. Hundreds of men were overcome as a cloud of gas rolled back in no-man's-land to engulf the King's Own Scottish Borderers. Showing unbelievable courage, the regimental piper, Piper

Laidlaw, pulled off his gas mask, grasped his bagpipes, and piped his men forward.

On another part of the front, Sergeant Packham of the Royal Sussex Regiment experienced the dreadful effects of the chlorine gas. Sent with a message to his platoon officer, Packham found the gas officer in the trench. "He looked ghastly and all the buttons on his tunic were green as if they were mouldy. He was saying that the gas was blowing back into our troops' faces. The wind had turned round on us."

The worst British casualties were suffered on the left wing of the attack by the Fourth Corps, notably by the First and Second Brigades of First Division. On this front the British line ran toward the northeast and with a wind blowing from south-southwest, it meant that here the troops were not only hit by their own gas but by that released by units to their right. Within a few minutes, over three hundred men of the First Brigade were down with gas poisoning. If possible, the Second Brigade was suffering even more. It was soon apparent that the British respirators were not working properly, and many men were gassed even though they were wearing them. Ironically, while the British gas weapon was inflicting heavy friendly casualties, traditional blade and bullet were helping the British soldiers to achieve a considerable success on the first day of the battle. Gas had played no part in this success other than in its capacity to conceal the advance of the British troops across no-man's-land. Smoke alone would have done as well, and without poisoning large numbers of friendly troops in the process. Haig's decision to release the gas in the circumstances that prevailed on September 25 was unjustifiable, and the heavy gas casualties of over two and half thousand men could have been avoided.

The Somme

One of the main aims of the British Fourth Army's offensive on the Somme in July 1916 was to seize possession of the Poziéres Ridge, which overlooked the British front lines. After the failure of the great July 1 assault, Douglas Haig made several other attempts to take the ridge, calling up the new Anzac divisions that had been brought to France after the collapse of the Gallipoli campaign. The Anzacs—Australians and New Zealanders—were outstanding troops and were eventually to form the main strike force of the British army in 1917 and 1918, but at this early stage they were relatively inexperienced in the techniques of trench warfare. Their initiation—at Fromelles and at Poziéres—was to see them subjected to a series of appalling friendly fire incidents as a result of poor coordination with their artillery.

At dawn on July 19, the Australians and New Zealanders began their attack on Poziéres after a period of two days' preliminary bombardment. However, the German artillery had not been silenced, and the Anzacs found themselves under fire from both in front and behind as their own artillery was firing short. Many Australian troops fell in no-man's-land as a result of short shells. In some areas, aware of the danger of hitting their own troops, the British artillery simply ceased fire, allowing the German machine gunners at the Sugar-loaf an opportunity to cut a swath through the attacking troops. As Brigadier-General "Pompey" Elliott later explained, the gunners "were really afraid, as we learnt later on, of aiming at the Sugar-loaf at all for fear of hitting our own lines." The truth was much starker than this. It was not only a case of the gunners not being willing to risk hitting their own men, but rather that they knew that their equipment was obsolete and that their skill was insufficient. The guns supporting this Australian attack in their assault were of the poorest quality, many quite antiquated, with a tendency to inaccurate fire. The gunners simply could not provide the precise creeping barrage that the infantry needed, without spraying shells into their own lines as happened all too frequently at this stage of the war.

The officer commanding the artillery for the Anzac Corps, Brigadier General Cunliffe Owen, bore the brunt of General Haig's dissatisfaction with the Australian assault. Haig sacked him on the spot. Only later did the truth emerge that Cunliffe-Owen had planned one barrage so badly that had it been fired, it would have achieved one of the most complete blue-on-blues in recorded history. The fact that it was not fired was entirely by chance. General Brudenell White, concerned about the haphazard British planning, chose to look in at corps artillery headquarters just as the gunners were being given their orders to fire. A quick calculation told him all he needed to know, and that particular barrage was stopped. Nevertheless, many Australians did suffer from friendly fire, and the Germans later reported that some of their Australian prisoners not only had endured fire from their own gunners but had also been fired at in error by British infantry.

As the British demonstrated by their use of gas at Loos in September 1915, some weapons could be as dangerous to friendly troops as they were to the enemy. A year later, the British were to find that this held true for tanks as well as for gas. The appearance of the first tanks on the Somme battlefield in September 1916 signaled one of the most important advances in warfare since the discovery of gunpowder. Yet in his hurry to use them, General Haig was willing to unveil this secret weapon before it was ready. Winston Churchill, for one, believed the tank should have been kept secret until Britain had enough ready to make a decisive breakthrough in the German lines. Lieutenant Colonel E. D. Swinton, the man most associated with the development of the early tanks, agreed:

> Some of the machines were asked to force their way through a wood and knock down trees—tricks which they had not been designed to play and which were likely to damage them seriously. I protested against these ''stunts'' and the frequent exhibitions, which were wearing out both machines and personnel. In addition to the almost continuous work of repairing, cleaning and tuning their Tanks, the men barely had time to eat, sleep and tend to themselves. I speculated as to how many machines would be one hundred per cent fit to go into action when their day arrived; and wondered how the Royal Flying Corps would have fared if it had made its début during the War with fifty aeroplanes of the first type produced, and had to submit to similar preliminaries before it went into action. As had been the case in England, it seemed impossible to establish a realization of the fact that the New Arm was a mass of complicated, and in some ways, delicate, machinery in an embryonic shape, and not the fool-proof product of long trial and experience.[18]

But Haig was adamant: The tanks must lead the assault at whatever cost, and as usual it was the common soldier who suffered.

The first tanks were really just armored machine-gun or light artillery carriers, with the ability to resist anything less than a direct hit from an artillery shell and able to cross trenches, crush barbed wire, and engage enemy pillboxes. In the context of the static fighting that had dominated the Western Front for two years, the generals might be excused for seeing in them the answer to all their prayers. However, the tanks had severe shortcomings, some of which imposed crippling limitations on the infantry that advanced alongside them. Frankly, the Mark I tank was merely a prototype and should not have been committed to action so readily. In the first place, it was painfully slow, capable of just fifty-five yards per minute (about two miles an hour), and was far slower than the infantry. Secondly, the tanks were so prone to mechanical failure that two of the six demonstrated to Sir Henry Rawlinson in the summer of 1916 broke down in ideal conditions and behind the lines. The sheer size of the tank and the fact that its slow speed made it an almost stationary target rendered it very vulnerable to artillery fire. Its weight also made it liable to bog down in the heavy mud. But what was worse from the point of view of the long-suffering infantry was that with a poor range of vision from within the tanks, their generally inexperienced crews were sometimes unable to differentiate friend from foe. As a result, tanks caused a number of friendly fire incidents, two of which were especially costly on September 15.

In spite of the tanks' limitations, the Fourth Army commander, Sir Henry Rawlinson was "on the whole . . . rather favorably impressed"—so much so that he welcomed the chance offered by Sir Douglas Haig to use the fifty tanks that were, at that stage, the only such weapons in existence. Rawlinson gave the tanks a key role in the attack on September 15, aiming—somewhat illogically—for the tanks to attack at night and move ahead of the British infantry to subdue the German strong points and wipe out the machine gunners where possible. In his view, the tanks would reduce the infantry casualties to a minimum, yet how he expected the tank crew to navigate even by the light of the moon was not made clear. In the end, the night option was canceled and an attack at dawn substituted. Obviously, Rawlinson was unaware of the tanks' real limitations, notably their slow speed and the poor visibility offered to the crew within. These two factors were soon to be the cause of heavy British losses.

The next major problem for Sir Henry Rawlinson in planning his new offensive was how to coordinate the tanks with their artillery supports. The assumption was that the tanks would move ahead of the infantry and eliminate obstacles and German machine gunners, but if this was so, it would clearly be impossible for the artillery to provide a creeping barrage for the infantry without hitting the tanks. So Rawlinson decided that the gunners would leave corridors in the barrage, within which the tanks could operate. Yet with corridors one hundred yards wide, there would be large sections—and the most formidable sections at that—of the German lines free of bombardment. Unless the tanks were able to subdue these sections, the infantry following them would simply be massacred. Rawlinson was expecting too much of the tanks, which could only engage German strong points from relatively close range, meaning that the British infantry would be exposed to German fire for much of their advance before the tanks could become really effective. And bearing in mind the snaillike speed of the tanks, as well as their vulnerability to gunfire, there was every chance that the infantry would overtake them and have to assault strong German defenses without any artillery support whatsoever. In this way, the appearance of the tank on September 15, rather than providing protection for the British infantry, actually contributed to even heavier casualties than would have been inflicted without them.

The pre-assault barrage of the German defenses began on September 12, and for three days the great guns cratered and plowed up the land, making it as unsuitable for tank warfare as can possibly be imagined. Combined with heavy rain, the British artillery now turned that area of the Somme offensive into a morass. By the fifteenth, when the attack began, the weather had improved, but the ground was still difficult for the tanks, and their top speed was reduced even more. It soon became impossible for the infantry to stay behind the tanks, and so they were forced to brave the German fire unprotected. In one area—the Quadrilateral near Bouleaux Wood—thirteen of the fifteen tanks allocated simply failed to turn up, getting lost on the way or breaking down. Here the British Sixth Division, ordered to attack the Quadrilateral head-on, was simply massacred by the German machine guns, while the Fifty-sixth Brigade and Guards Division were enfiladed by machine-gun fire. At one stage, a group of men from the Worcesters were mistaken for Germans and shot down by a British Vickers machine gun.

Few infantry commanders of the time understood the difficulties faced by tank crews. Inside the steel ''coffins,'' the tankers had very limited vision and were constantly deafened by the unbearable noise of the 105-horsepower gas engine. In addition, the heat and fumes were overpowering for the eight-man crew, who had somehow to navigate the cumbersome machine through the most difficult terrain ever devised for man as a battlefield, while suffering the discomfort of near-misses by German heavy artillery and the constant rattle of machine-gun bullets striking their armored exterior. In such conditions, it was hardly surprising that tanks frequently lost their way and inflicted friendly casualties. Even before reaching the start line on September 15, one tank had already fired its six-pounder main armament at the battalion headquarters of the Post Office Rifles, while another nearly succeeded in destroying the headquarters of the Civil Service Rifles. Navigation was so difficult that tank commanders frequently stopped their tanks and got out to ask the infantry which way they were supposed to be going.

Incredibly, September 15 was not a day of total failure. Three divisions of the Fifteenth Corps, attacking toward Delville Wood and Longueval, achieved the greatest success, taking the village of Flers and overrunning the second line of German trenches. Here two tanks were responsible for the breakthrough, with the Germans showing little fight and surrendering, many in terror of the new weapons. But elsewhere, the tanks had played a more deadly part in the proceedings. The idiotic General Pulteney, commanding the Third Corps, had widened his artillery corridor on the assumption that his tanks would quickly capture the German positions in High

Wood, and therefore his infantry would not need a creeping barrage. In fact, the tanks were quite unable to operate among the trees, with the result that the Germans were free to bring a heavy fire down on the infantry of the Forty-seventh Division. Four tanks gallantly tried to move through High Wood on Pulteney's orders, but immediately lost their bearings in the confusion of tree stumps and German shells. Two of the tanks turned completely around without noticing it and emerged from the wood into British lines. One became bogged down in a shell hole, but the other, commanded by Lieutenant Robinson, seeing trenches filled with troops, assumed that it had reached the German positions in High Wood and immediately raked the trenches with machine-gun fire. But the trench was a British one, known as Worcester Trench, and the men were from the London Regiment. In a storm of fire, dozens of men were killed and wounded before the tank commander realized his mistake. The immediate blame for this blue-on-blue was his, and yet the greater error was made by General Pulteney in ordering tanks to clear a wood without artillery support. Eventually, High Wood was taken, but only after a heavy mortar bombardment paved the way for an attack by New Zealand infantry.

There was great excitement at Fourth Army headquarters when it was reported that two tanks had broken through the German lines and reached the village of Flers. General Haig even passed the news on to the waiting pressmen, and it was relayed back to Britain. What was not reported was that another tank had wiped out nearly all the troops in a nearby trench—a British assembly trench. The Ninth Norfolks had been in the trench, preparing to go over the top when a tank lumbered up, lost its bearings, and confused their trench with the German front line. At once its machine gun raked the trench, killing many of the helpless soldiers. An infantry captain, Captain Crosse, ran toward the tank, waving his arms and trying to make himself understood. Peering through a slit in the armored side, the machine gunner at last understood, and the tank stopped firing and swung away, having wrecked the British attack and leaving the German front line unharmed. When the remnants of the Norfolks attacked later that day, they were cut down by the same German defenders who had been the tank's real target.

The first appearance of tanks in battle on September 15, 1916, had not been an unmixed blessing for Sir Henry Rawlinson. It had been expected that the tanks would not only help in overrunning German trenches and capturing the target village of Flers, but would also help to keep casualties to a minimum. In two separate ways, the tanks did the very opposite. In two serious cases of amicide, they had killed many British soldiers, while in their failure to keep up with the advancing infantry, they had contributed to the massacre of thousands of men by unhampered German machine guns and strong points. July 1, 1916, is known to history as the black day of the British army, during which over sixty thousand casualties were suffered. What is not remembered, however, is that on September 15, 1916, British Fourth Army casualties of 29,376 were suffered from an attacking force only half the size of the one that attacked on July 1. And yet, of course, both Haig and Rawlinson—and the British public—viewed the September 15 offensive as a success and the performance of the tanks as a triumph. This overlooked the fact that in 1916 the tank was a luxury weapon that, because of the high price it exacted in friendly casualties, few armies could afford.

As the Battle of the Somme continued into the fall of 1916, the British struggle to capture and hold the strategic High Wood was accompanied by a number of important blue-on-blue incidents, mostly involving the artillery. During an operation by the Royal Engineers to tunnel under the German defensive redoubt, the Glasgow Highlanders were hit by a series of short barrages from British guns that cost them three officers and forty-five men dead and wounded. In the days just before the August 18 Anglo-French assault, the British field artillery worked to perfect their creeping barrage. Their aim was to lay down a curtain of fire some one hundred yards in front of the advancing infantry. One minute later, during which time the gunners calculated the foot soldiers would have moved forward fifty yards, the barrage would be lifted a further fifty yards, and so on every minute. This kind of precision, while possible in ideal conditions, was quite beyond gunners who were themselves subject to counter-battery fire and who were using artillery pieces that had grown worn and unreliable from heavy use. And while the gunners suffered such heavy casualties themselves, new and inexperienced crews were constantly being brought in who rarely survived long enough to perfect the techniques necessary to run an efficient creeping barrage. The result was that the gunners lost confidence in themselves and in their guns, while the infantry lost confidence in the gunners. As one gunner told an infantry officer: "If we fire over you, God help you—we've only one trained gunner per gun left." The gunner was prescient: The following day the Second Battalion of the Argyll and Sutherland Highlanders was caught by its own barrage near High Wood. The British shells "walked through" the Highlanders' positions, causing heavy casualties. To make matters worse, the Scotsmen had with them some new gadgets, known as pipe-pushers, which were designed to speed the entrenching process for assault troops. Unfortunately, when the Scots tried to use them they simply blew up, one blowing such a crater in their position that it exposed many of the Scottish troops to sniper fire.

The Battle of Passchendaele

One of the keenest critics of the British army in the First World War was the Australian military correspondent and later official historian, C. E. W. Bean. Bean would have been the first to admit his partiality when it came to discussing the performance of British generals, selected as most of them were from a system that favored men who had attended certain privileged schools and that was above all a reflection of the rigid class structure he saw as typical of pre–1914 Britain. On the other hand, his honesty and independence allowed him to write about things that the British would rather have concealed. On the question of friendly fire, for example, the historian owes much to Bean's incisive pen, for he will seek in vain to find friendly fire given much coverage in the official histories. As Bean has made clear, throughout the conflict the British artillery all too often fired short in support of the infantry, and, from 1916, those who suffered involved his own beloved Australians.

It has been said with much justice that the First World War was an artillery war and the big guns did most of the killing and the maiming—some estimates claiming that nearly three-quarters of all wounds were caused by artillery shells and shrapnel. But it was also the sheer violence of the guns that administered the "shell shock" to so many soldiers. Yet the art of long-distance "indirect" artillery fire was still in its infancy, and the precision needed to achieve a perfect creeping barrage to precede an infantry attack was a skill few gunners on the Allied side could boast even in 1917. And it was in trying to provide these creeping barrages that the British and French guns killed and wounded so many of their men. On the Western Front, the Germans had adopted a defensive posture since 1915, and it was not until the spring of 1918 that the Germans returned to the offensive, strengthened by troops returning from the east, where Russia had been defeated. Thus for three years of the war, the Germans were free of much of the short firing that, according to Percin, cost the French so many casualties and that, it must be conceded, may well have cost the British almost as many. Without a Percin to provide figures, one can rely only on scattered reports in diaries and memoirs, or use the reports of a keen student of military tactics like Bean. The incidents are scattered across miles of battlefront and follow no sort of pattern. Yet they are not mere isolated examples. Almost every British soldier would have had a fund of friendly fire stories, and most of them would be at the expense of the gunners, the unseen killers who killed at a distance and never saw their victims.

The PBI (poor bloody infantry) were not always passive receivers of short firing by the artillery. During the Second Battle of Ypres in May 1915, Corporal A. Wilson of the West Yorkshire Regiment intended to find out the truth about friendly fire. With his friend Walter Malthouse, Wilson was in the trenches near Fauquissart when a shell suddenly landed alongside them:

> I was stunned, of course, but when I got my wits together I could hardly believe it, I was covered in blood—saturated—

> and I really thought I'd bought it. But it was Walter's blood. I didn't have a scratch myself. Walter had taken the full blast and somehow or other it hadn't touched me, He was blown to bits. A terrible sight. I don't think there was a bit of his body bigger than a leg of lamb. I gathered up what I could, put him into a sandbag and later on when it got to dusk, a few of us got out of the trench and buried him . . . The worst of it was that from the direction of the shell we felt almost sure it was one of ours. Of course the authorities wouldn't have that! But I got a few of the men digging and it only took us half an hour to find the nose-cap. Sure enough it was marked "WD"—War Department. It was from a naval shell fired by one of our long-distance guns mounted on an armored train.[19]

During the later stages of the Battle of Messines on June 7, 1917, the problem of artillery supporting advanced troop positions was starkly illustrated by a series of blue-on-blue incidents. Australian troops of the Forty-seventh and Thirty-seventh Battalions were holding a position known as Hun's House, southwest of Messines. German and British troops were closely engaged, and when a British aircraft flew over to spot for artillery support, the Australians, fearful of revealing their positions to the Germans, refused to fire their identification flares. The result was that the gunners at the rear were unaware of the fact that they had advanced so far. One Australian named Shang, situated some four hundred yards behind the most advanced troops, managed to signal his position by a Lucas lamp, and his position was assumed to be that of the most advanced British troops. When, therefore, a German counterattack was observed, the artillery prepared to intervene. Meanwhile, at Hun's House, the Australians, though heavily outnumbered, fought off the German attack, only to find themselves under a British bombardment. Assuming at first that the shells were German, the Australians fought on, until suddenly they were hit by the full fury of the British barrage. As Bean writes: "Their position was deluged with shell. Roots were torn from the hedge and tossed in the air; shrapnel began to crack overhead. A tree split and crashed. Fragments of steel swished along the ground and lay smoking." A wounded officer checked an attempt to retire, but eventually the men fell back to a line being established by New

Zealand troops. Yet going back meant once again passing through the British barrage, and more men were killed in the process. Eventually, with their men having retired, just two officers—Captain Williams of the Forty-seventh and Captain Allen of the Forty-fifth—stood facing the Germans together. As Williams later explained, his men "would stand all the enemy fire you like to give them, but they would not put up with being shelled by their own guns." Some of the Australians who had fallen back now took up positions at the "Owl Trench" on an earlier line at Oosttaverne, but the majority retreated all the way back to where the New Zealanders were still digging.

The retreat of the Australians naturally came as a shock to the New Zealanders, who were still consolidating their position. As a result, some of the New Zealand officers, assuming that there were no longer any Australian troops beyond their line and fearing a German attack, asked the gunners to shorten their barrage and bombard the Oosttaverne Line. This was done, and soon the Australians who were grimly hanging on to Owl Trench were driven out once again by their own guns. Not surprisingly, morale among the Australian troops was becoming very fragile. Part of the Thirty-seventh Battalion, however, still held on to the line of the Owl Trench, but again, through faulty coordination, the gunners assumed that all the Australians had withdrawn, and so the barrage was once again shortened to prevent a German follow-up. This new barrage fell on the doughty members of the Thirty-seventh who were still hanging on, and they were in turn forced to retreat, through their own barrage. The entire day's fighting on this front was a fiasco. Bean later wrote: "Thus, owing to the action of its own artillery—for which defects in the maps, over-eagerness of the infantry, over-anxiety of some of the staffs and commanders, and a dangerous degree of inaccuracy in the barrage was responsible—the whole of the final objective between the Blauwepoortbeek valley and the Douve had by 9 p.m. been left open to the enemy."

North of Huns Walk, the Australians were encountering similar problems with British artillery. At 0805 hours on June 7, word spread that the Germans were planning a major counterattack, and Captain Maxwell of the Fifty-second Battalion called in artillery fire to break up a German advance. Unfortunately, the gunners were firing short and deluged British positions behind the front lines. In a rear strong point held by the Sherwood Foresters, a Royal Engineers officer was killed, while the entire unit from the Sixth Border Regiment was driven back from Van Hove Farm. Maxwell, alarmed at the devastating effects of the friendly fire and failing to get it stopped by signal or messenger, set off himself to stop the gunners. At battalion headquarters Maxwell heard to his surprise commands being issued: "Load! Fire! See them on the right there!" As Bean writes:

> Rushing forward with a furious question as to what was going on, Maxwell found himself facing a British battalion commander, who said that the line had fallen back and that he was directing fire on the advancing enemy. The young Tasmanian offered to go down to this target himself to prove that the men seen were not German; immediately afterwards a flare revealed them, and a patrol found them to be a party of British machine-gunners searching behind the lines for a new position.

But so powerful were the effects of the rumors that were rife that day of a major German counterattack that the gunners were firing at anything coming from the direction of the German lines, even their own troops falling back.

The Third Battle of Ypres—popularly known as the Battle of Passchendaele—has a very fair claim to being considered the most terrible battle ever fought in military history. It is difficult to accept the idea of man as an intelligent animal when one considers this grimmest of slaughters, conducted in what historians were quick to describe in terms of John Bunyan's "Slough of Despond." Yet the battle's architects, Sir Douglas Haig and Sir Hubert Gough, never lost faith, and in spite of enormous losses in manpower, equipment, and, most of all, morale, they continued to believe that something worthwhile could be snatched from the Germans to justify all the bloodletting. And so the British kept the battle going, from July into October and November 1917, with the aim of taking Passchendaele village itself. To do this, Haig decided to use his best strike troops—the First and Second Anzac Corps.

Their initial target was to take the village by October 12, after which the Canadians and the British Cavalry Corps would push on to the railway junction at Roulers and cut the Germans off from the coast. It sounded easy, but the condition of the ground made it simply impossible.

For what history records as the battle of Poelcapelle, two British divisions—the Forty-ninth and Sixty-sixth—were earmarked to carry out preliminary attacks to pave the way for the Australians. The terrible fate of the Sixty-sixth Division constitutes one of warfare's most horrible examples of blue-on-blue. In the first place, the choice of the Sixty-sixth for so arduous a task was a particularly bad one. The division had no previous battle experience and had only been in France for a few months. In view of its inexperience and the apparent incompetence of many of its officers, it was assigned an area of attack where British intelligence reported that there was no barbed wire. They were mistaken. In addition, ground conditions were so appalling that it was decided to bring the men up to an assembly point just two and a half miles from their starting tapes. They would therefore begin their march to the front line at 1900 and, assuming at worst a speed of half a mile an hour in the thick mud, they should arrive at the front line by midnight, allowing them five hours' rest before the 0520 attack. That, at least, was the general idea. But with nine thousand men trying to march forward in darkness across muddy and slippery duckboards, chaos was never far away. Lieutenant Patrick King of the East Lancashires reported his experiences:

> It was an absolute nightmare. Often we would have to stop and wait for up to half an hour, because all the time the duckboards were being blown up and men being blown off the track or simply slipping off—because we were all in full marching order with gas-masks and rifles, and some were carrying machine-guns and extra ammunition. We were all carrying equipment of some kind, and all had empty sandbags tucked down our backs.[20]

The movement of the heavy guns over ground already possessing the consistency of porridge made the forward progress of the troops almost imperceptible. Every step was a battle. Men fell off the boards and were at once swallowed up in flooded shell holes. Dozens drowned in the darkness, unable to get free even with the help of colleagues. To make matters worse, every man was loaded with sixty pounds of kit and weapons, ensuring that they could drown even in a few feet of water, which, in the words of Leon Wolff, "was foul with decaying equipment, excrement, and perhaps something dead; or its surface might be covered with old, sour mustard gas. It was not uncommon for a man to vomit while being extracted from something like this." In places, the men of the Sixty-sixth were trying to march through liquid mud more than three feet deep.

At midnight—by which time they should have reached the frontline trenches—few of the men had covered as much as a mile. Even the deadline of 0500 seemed impossible now, and the decision was taken to get as many men as possible forward, with the rest joining in the attack as soon as they arrived. Incredibly, many men of the Sixty-sixth—green troops that they were—kept marching in darkness, high winds, and torrential rain, and under a heavy German bombardment, reached the front after a ten-hour march, fixed their bayonets and marched straight out into no-man's-land. There they were shattered and slaughtered by their own guns. Lieutenant King was one of the men who arrived in time to help organize the attack:

> The Colonel had led the battalion up the track—Colonel Whitehead, a very terse man, a very brave man. And he said to me, "Get them into the attack." I passed it on to the NCOs, who gave the orders: "Fix bayonets. Deploy. Extended order. Advance!" We went over into this morass, straight into a curtain of rain and mist and shells, for we were caught between the two barrages.[21]

King watched the attack disintegrate under machine-gun fire from the Germans and short artillery fire from the British guns. With a handful of men, he occupied a shell hole and sandbagged the rim. Here he had to remain for twenty-four hours in heavy rain and under constant shelling from both sides. He does not seem to have felt any bitterness toward the British gunners who, by this stage, had no real way of knowing where the British front line was. They seemed to be lobbing shells hopefully toward the Germans, but most were falling among the wretched groups of the Sixty-sixth Division, as they lay scattered in shell holes in no-man's land. The other British division—the Forty-ninth—fared little better, having to cross the flooded Ravebeck River, a fifty-yard-wide steady stream of waist-deep liquid mud.

At dawn on October 10, Lieutenant King, in his shell hole, was surprised to hear voices behind him. "Who the hell are you?" he called out to a group of unidentified soldiers. "Well, come to that, who are you?" replied one of the men. "I'm Lieutenant King of the 2/5th East Lancashire Regiment." "Well, we're the Aussies, chum" was the reply, "and we've come to relieve you." At which, to King's astonishment, the Australians proceeded to occupy his shell hole, already half-filled with putrid water. As King and men like him struggled back to the British lines, Charles Bean recorded the views of one of them who had been injured in the attack:

> Ah doan' know what our brigade was doin' to put us in after a twelve hours' march—twelve hours from beginning to end. We had no duckboards like these—we plugged through the mud. We didn't know where the tapes were, and by the time we arrived there our barrage had gone on half an hour. The men were so done they could hardly stand oop an' hold a rifle. We didn't know where our starting position was, but we went on after the barrage. I'm sorry for the Australians, and it was our first stoont too. We're a new division, ye know.[22]

The truth was that the men of the Sixty-sixth Division, who had attacked at dawn the previous day, had been so thoroughly exhausted that they had fallen far behind the creeping barrage that had been arranged for them. Their divisional commander, Major General Sir Henry Lawrence, therefore ordered the British gunners to bring back the barrage, which, without checking where the Sixty-sixth had gotten to, the artillery did by simply shortening its fire. The result was that British shells cut the Sixty-sixth Division to pieces in no-man's-land, inflicting heavy casualties. The massacre was witnessed by Australian troops, one of whom reported what he had seen to the Australian commander, General Birdwood. Birdwood at once complained to GHQ that he had just heard that British soldiers had been wiped out by their own gunners. But the commander of the Sixty-sixth Division, no doubt embarrassed rather than concerned over the blue-on-blue, furiously criticized Birdwood for reporting this tragedy and demanded to know where he had gotten his information. Birdwood had, in fact, been informed by an Australian officer who had personally witnessed the carnage. Lawrence then insisted that Birdwood withdraw his charge of amicide and discipline his witness, or else he would personally block any promotions from within the Australian Corps. The matter was eventually hushed up, with Lawrence being promoted to the position of Sir Douglas Haig's chief of staff, as a reward for his ineptitude and dishonesty. It was obvious that the British were less willing to acknowledge the existence of friendly fire than their French allies.

Meanwhile, General Douglas Haig was confident that the Australians could achieve their mission, which was to take Passchendaele village. "The New Zealand and Australian 3rd Division are to put the Australian flag in the church there," he told his wife. But there were no flags and no celebrations; the attack was a disaster, with the Australians suffering 60 percent casualties. Even Haig was now forced to admit that the ground was "quite impossible."

But still the battle went on. Now it was the turn of the Canadians. On October 16, Lieutenant General Currie, commander of the Canadian Corps, declared that if the weather improved and enough artillery could be assembled in time, he was prepared to commit his troops to a new attack on the Twenty-sixth, followed—if necessary—by further assaults on October 29 and November 2. The important provisos should be noted: if the weather improved and if there were enough guns. In fact, for a while in the middle of October, the weather did relent. But the Canadians discovered that fewer than half of the British guns that had been promised would be available; the rest were under water or clogged with mud or just "missing." They were also too far back and generally out of communication with each other. When the Canadian commander complained and demanded more guns, he was told that he would have to send an indent to GHQ, as there were no more available.

However, it is all too easy to criticize the much-maligned British gunners. With the rain incessant through October, one gunner reported that in his battery the water was as deep as the guns' breaches, which went under water every time the guns were fired. The gunners had tried to convince GHQ that if the battle went on, there would be no heavy guns at all for use in 1918, as they would all be thoroughly worn out. Colonel Rawlins, artillery adviser to Haig, even had the temerity to make this view known to the great man. Haig went white and shouted, "Colonel Rawlins, leave the room!" When Brigadier-General Edmonds tried to support Rawlins, Haig turned on him: "You go too, Edmonds!"

Lieutenant Colonel Alan Brooke, of Second World War distinction, but at that time attached to the Canadians as the chief artillery staff officer, attended one of Haig's briefings for the Canadians and wrote later that he could hardly believe what he heard on that occasion. Haig spoke as if the attack was taking place in normal conditions against a weakening enemy, on the point of rout. Brooke suggested that Haig could never have seen what it was like at the front. In fact, the British commander had closed his mind to all objections.

The Canadians began their attack at 0540 hours—in heavy rain. Nothing had really changed from when the Australians failed a fortnight before and yet German resistance was weaker, and by the end of the day, the Canadians had made progress and were occupying land to the west of Passchendaele itself. But to reach them with ammunition and supplies was almost impossible through the mud, and hundreds of mules died by drowning after slipping, fully laden, from the greasy duckboards that were hastily brought up to bridge the morass. Two days passed, and the Canadians tried again, inching closer to their goal. On October 30, the Germans began to give way, but the advance was costing the Canadians up to fifty percent casualties. Five German counterattacks were repulsed. Now that the Canadians had gotten a foothold on the high ground of the Passchendaele Ridge, they were not going to allow themselves to be driven back into the muddy horror through which they had fought at such cost. On November 6, the village of Passchendaele that Douglas Haig had valued at a third of a million British casualties at last fell to the Canadians, and a week later, the battle was brought to a close. But the last dismal fortnight of the fighting, which had brought such success to the Canadians, had contained at least one serious blue-on-blue incident. During the assault of November 6, and as a result of inaccurate orders, two companies of the Canadians had been positioned one hundred yards ahead of instead of behind the artillery barrage. As the creeping barrage advanced, both companies were blown to pieces by the British guns. The Canadian survivors, retreating from the slaughter, were then attacked in error by two British companies supposedly supporting them.

Denis Winter in his recent book, *Haig's Command*, has described the reaction of Canadian Prime Minister Sir Robert Borden to what the Canadians had had to endure. It so infuriated Borden that at one meeting of the imperial war cabinet, he took British Prime Minister David Lloyd George by the lapels and shook him furiously.

For all the titanic efforts the British had made during the fall of 1917, Haig had merely demonstrated his—and Britain's—incapacity to win the war alone. Britain would now have to "sit tight and wait for the Americans" to redress the balance on the Western Front. The Battle of Passchendaele, for all its secret political agenda, had been what everyone believed all along, a doomed offensive that should have been aborted as soon as its faint trickle of life began to fade, just days into the offensive.

Even during the victorious Second Battle of Villers-Brettoneux in August 1918, the Australians were still having trouble with the British gunners who were supposed to be supporting them. The Australian Fifty-third Brigade came under fire from British guns, presumably as a result of an aircraft reporting the wrong grid references to the gunners. Sergeant Kahn of the Fifty-fourth Brigade had been responsible for running the telephone lines between the gunners and the brigade headquarters. Seeing the friendly fire developing, he desperately tried to stop the guns firing, only to find that the telephone lines had been cut by enemy fire. Kahn and his colleague, Sergeant Sheppard, found the break and managed to mend it. Sheppard was at once in touch with his brigade headquarters. "For God's sake stop shooting," said Sheppard, "You're firing on our own men." "Who is speaking? What authority have you for stopping the guns?" was the reply. "Linesman here," said Sheppard, "If you will hang on a moment we will have the line mended quickly and then you can speak to an officer." Kahn and Sheppard had risked their lives, running out under the British barrage to find the broken lines, but they had saved far worse casualties. Although five men were already dead and seven wounded, Kahn commented that the guns had just adjusted their range and would soon have been putting shells right into the Australian dugouts, with an appalling slaughter.

THE SECOND WORLD WAR

The factors that returned mobility to the battlefield in 1918, namely tanks and aircraft, were also those that offered the greatest threats of amicide in World War II. With far more fluid combat situations and with operations stretching across a much vaster area, the problems of locating and maintaining contact with friendly forces became very great. With fast-moving armored formations pushing the battlefront back at speed, there was always the threat of friendly casualties from indirect artillery fire and tactical ground support aircraft. In spite of improved communications, errors were frequent. And the difficulties of terrain added substantially to the problems of identification. The fighting in Europe, notably in Italy in 1943 and in Normandy in 1944, frequently involved house-to-house fighting, sometimes amid rubble that could conceal both friendly and enemy troops. In the North African desert, with its absence of landmarks and identifying features, where sandstorms and dust concealed vehicle identification, mistakes were frequent. Perhaps worst of all was the jungle fighting in Burma and New Guinea, where it was often impossible to follow the movements of friendly units and where tactical air support had at best to be something of a hit-or-miss affair.

The performance of inexperienced American soldiers in

North Africa in 1943 has been well-documented, as have incidents of friendly fire among the green troops. Undertrained and too soft, the GIs seemed fueled only by a kind of naive faith that things would work out in the end. One American officer, apparently aware of the fragility of his company's morale, wrote:

> We never really thought we were hot in camp. We always thought that out in the theater they would do it better—the sentry would never daydream, the communications would always work and our units would always be where and when they were supposed to be—we could see a hundred flaws in everything we did, a hundred ways in which an alert enemy could beat us by capitalizing on our errors. Sometimes we made big talk about what our unit could do in action but we always knew it was just that.[23]

During the ill-fated airborne operation over Sicily on July 10–11, 1943, Charles Shrader has shown that even when the paratroopers landed they were not safe from the friendly fire of American ground troops. Landing well to the southeast of Gela, Chaplain Kuehl and a small group of the 504th were immediately brought under fire by friendly forces. Even though they knew the password and yelled it at their assailants, this did not stop the fire, and it was only after the chaplain had crawled to the rear of his assailants and reasoned with them that they stopped firing. On the same night—July 11—the American 171st and 158th Field Artillery Battalions were in an action against Colonel Reuben Tucker's paratroopers. As one of the gunners pointed out quite reasonably: "Since no news of the American paratroopers had reached this headquarters, they were assumed to be hostile and the Battalion was deployed for all round defense." The outcome of this engagement was that one of the paratroopers was killed. Clearly, as we have seen elsewhere (see page 150), coordination among the three armed services during Operation Husky reached an all-time low. On the other hand, the ground troops appear to have shown a complete absence of initiative. While the gunners shot the easily identified C–47s out of the sky without a thought, they also kept up a steady fire against men who called out to them in their own language and supplied the right password. In view of the overwhelming evidence that a mistake had been made, American troops continued to fire at and kill their own men without anyone showing the initiative to call a halt or radio to their headquarters that they had doubts about the identity of the troops they were engaging. Certainly night added to the fears of the individuals, and the strategic situation, namely that they had just landed on an enemy-held island, must have contributed to their unease, yet these soldiers should have been more than unthinking automatons. The men on the destroyer *Beatty*, who kept firing at a ditched C–47, which was clearly identifiable, and the tank crews who machine-gunned the paratroopers in their harnesses, had succumbed to the primitive urge to kill in which conscious thought played no part. In such situations, friendly fire is an inevitable consequence.

During the fighting around Salerno, in September 1943, green American troops reacted badly to rumors. Norman Lewis, attached to Fifth Army headquarters, wrote of what he saw:

> Outright panic now started and spread among the American troops left behind. In the belief that our positions had been infiltrated by German infantry they began to shoot each other . . . Official history will in due time set to work to dress up this part of the action . . . with what dignity it can. What we saw was ineptitude and cowardice spreading down from the command and this resulted in chaos.[24]

The Second World War saw a transition in the use of artillery, from the massive, static bombardments seen on countless occasions on the Western Front in 1915–17, notably on the Somme in June 1916 and at Passchendaele in the summer of 1917, to a more mobile and much more flexible use of guns after 1940. The massive barrage still played a part in battles of this period, as for example by the British under General Montgomery at El Alamein in 1943, but the blitzkrieg tactics introduced by the Germans against Poland and France stressed mobility as well as firepower, and a battlefront could change so quickly, particularly on the Eastern Front, that immobile units of artillery would have been overrun and left miles behind the action. Combined with this increased mobility came the almost universal employment of indirect fire, requiring a precision in target identification that was difficult to achieve at the time. This resulted in many friendly fire incidents that have had to be accepted as a small but inevitable part of modern warfare. Charles Shrader has expressed surprise at the relatively small number of cases of artillery amicide reported from the North African theater of operations in 1942 and 1943. Here, green American troops were responsible for numerous examples of aerial and ground amicide, and it may be that there are fewer available sources than later incidents from which to trace artillery amicide instances. Or, indeed, blue-on-blues may have been so regular that they provoked little notice. For whatever reason, artillery incidents of amicide during the Second World War produced none of the truly horrific examples cited by French general Percin and British writers during World War I.

The role of the artillery during the Allied advance through Italy from 1943 to 1945 was significant. The campaign involved fighting both in mountainous areas and in heavily built-up urban areas. In both, German forces offered tenacious opposition, generally needing to be dug out of powerful defensive positions. Mobility was far less a factor than in North Africa or Normandy, and the attritional struggle was often far more like those typical of the First World War. The result was that artillery played the vital part in softening up the enemy prior to an assault, and in consequence friendly casualties were often heavy, notably where frontline troops became intermingled with those of the enemy in street fighting.

During American assaults on the Gothic Line at Monte Altuzzo in September 1944, a series of blue-on-blue incidents occurred. On four successive days, September 14–17, elements of the Eighty-fifth Division suffered casualties from American artillery on four separate occasions. Fighting on the steep and rocky slopes of Monte Altuzzo, men of the First Battalion, 338th Infantry, were hit by mortar fire and artillery shells, pinning them down and wrecking their assault. The Germans were threatening to break the American hold on the mountain by vigorous counterattacks, and the shorts fired by their own gunners forced Company B commander Captain Maurice Peabody to pull his men back to Paretaio. Even while engaged in withdrawing, the 338th was not safe from its own guns. Before they reached Paretaio, they were hit by American shells, which had been called in to support a neighboring company of the Second Battalion against a major German attack.

Meanwhile, Company C of the First Battalion was holding the southwest slopes of the mountain, and on September 15, they were also hit by incoming American fire. One shell struck a group of soldiers, killing six of them, including an officer, and wounding several others. The commanding officer of the First Platoon, Lieutenant William Corey, rightly guessing that the shells were American, tried to get the firing stopped, but found that his telephone lines had been cut in the explosion. When he did eventually get his message through to the supporting artillery—329th Field Artillery Battalion—they vehemently denied that the shells had come from them. In fact, the truth was never discovered. But as Charles Shrader has shown, this incident—minor indeed compared to the aerial holocausts of Operations Cobra and Totalize—was very significant in the effect it had on the morale of the infantrymen. As Shrader writes:

> Although only one shell actually struck the platoon, the resulting confusion and the belief that it was from friendly artillery had a demoralizing effect on the survivors. The dazed and angry men had quickly scattered down the hillside and Lieutenant Corey had an extremely difficult time reorganizing his position to face the expected German counterattacks. The shaky men were scarcely capable of maintaining a stubborn defense.[25]

A few misdirected shells had an effect out of all proportion to the casualties they inflicted or the damage they had caused. They had seriously weakened the morale of troops who were already at full stretch holding back a difficult enemy. For the American soldiers in this instance to feel that they were as vulnerable from behind as they were from the front was a blow to a fundamental aspect of warfare: security to the rear. As we saw in the case of General Braddock's British troops on the

Monongahela River in 1755, the belief that they were being encircled caused morale to collapse and panic to set in. On Monte Altuzzo, the effect of the few friendly casualties was that the infantry felt unwilling to call in or even accept mortar and artillery support, and without this support, infantry assaults were almost impossible. Later attacks by A and C companies of the 338th were carried out without support from their mortars for fear of having frontline troops hit by their own shells. In consequence, the German defenders were free from artillery bombardment and no doubt inflicted heavier casualties on the advancing Americans than would have been the case had they had to keep their heads down. And yet all the evidence suggests that soldiers would much prefer to risk enemy fire than incur a real—if much smaller—danger of being hit by their own shells. What was lost on Monte Altuzzo was not just artillery support, it was trust. And if soldiers within a unit—battalion, regiment, or division—cannot trust each other, and lose confidence in their supporting artillery, then not only does morale suffer but fighting efficiency as well.

Normandy and After

After D-Day, June 6, 1944, the historical record of the progress of the U.S. Thirtieth Infantry Division through the Normandy battlefields has provided Charles Shrader with much evidence of the problems of friendly fire in the European theater of the war. Examples of amicide were so common during the fighting of 1944–45 that it is only possible to deal with a few of the cases that affected this single division that, in a sense, can stand as an exemplar of what was happening generally in the American sector of operations.

On July 7, Lieutenant General Corlett, commanding the Nineteenth Corps, of which the Thirtieth Division formed a part, hoped to effect a breakout by using part of the Third Armored Division—the CCB—to cross the Vire River and attack southward, in conjunction with the Thirtieth Division. This decision was to set in train a series of confused events that led to incidents of amicide. When the tanks and armored vehicles of CCB reached the Vire River at Airel on the night of July 7–8, they found the whole area crowded with the Thirtieth Division infantry. In the milling confusion, it was difficult for them to get their heavy equipment across the river by the single bridge. In a startling modern rerun of the Austrian disaster at Karansebes in 1788, the frustration felt by the American troops caused them to begin firing in the air and then, shockingly, at each other in their frustration. Even machine guns were brought into action against their fellow Americans. The Thirtieth Division's commander, Major General Leland Hobbs, shocked by the previous night's madness, complained to corps commander Charles Corlett that his men had suffered sixteen casualties, all shot by members of the CCB. There were no casualties reported by the CCB, though the intensity of the firing makes it unlikely that they escaped unscathed.

Even when day came on July 8, the situation did not improve. It was hard to separate the thousands of infantrymen from the heavy equipment because there was no obvious place for them to reassemble once they had become crammed into the streets of the small Norman town. To add to the confusion, the whole area was under counterattack by German forces of the Second SS Panzer Division. General Hobbs tried to keep an even temper, which was difficult in view of the fact that forward units of the Thirtieth Division were in contact with German forces and could not be supported by the divisional artillery, which was trapped in the logjam in Airel. The only answer was to put the two units under the same command, and that is what Corlett decided to do, placing the CCB under Thirtieth Division command, not entirely to the satisfaction of General Hobbs, who had had a bellyful of the troublesome tankers.

By July 9, Hobbs may have felt that he was free at last of the problems of Airel, of friendly fire, and of the CCB when he was able to order the tankers to attack southwest and capture a vital strategic position known as Hill 291. But Hobbs was wrong. The unholy marriage of CCB and Thirtieth Infantry was just over the honeymoon period. The commander of CCB, Brigadier General Bohn, was having trouble moving his heavy vehicles through the Normandy bocage, the difficult hedgerow terrain that held up Anglo-American forces in the first weeks after the Normandy landings. To make matters worse, heavy rain had made the ground thick with mud. Even had he wished to press on and get out of the way of the Thirtieth Division, he was finding that the elements were against him. Eventually, everybody lost patience with what they saw as his delaying tactics. His new commander, Leland Hobbs, gave Bohn an ultimatum: Either get going and take Hill 91 by 1700 hours or get sacked. This spurred Bohn into action. He ordered one of his tank companies to cross the St. Jean de Daye–Pont Hebert Road and head straight for Hill 91 and not to stop for anything. At once, eight Sherman tanks set off toward Haut-Vents, firing as they went and spraying machine-gun bullets into the surrounding hedgerows. For the moment, they were out of the picture, but they would be back in time to steal the show in the last reel.

General Hobbs had other things on his mind besides the progress of Bohn's tanks. Thirtieth Division was about to become the meat in a sandwich, attacked by the Second SS Panzer Division from the west and the Panzer Lehr Division from the east. During a day of heavy fighting, Hobbs's attached armored unit, the 743rd Tank Battalion, was virtually wiped out in a German ambush. Hobbs's 823rd Tank Destroyer Battalion, armed with thirty-six three-inch and 76mm antitank guns, had also been under intense pressure from German thrusts. On the afternoon of July 9, First Platoon, Company C of the 823rd, led by Lieutenant Ellis McInnis, was in a defensive position south of the St. Jean de Daye crossroads. Retreating infantrymen reported as they passed him that German tanks were not far behind, and shellbursts overhead indicated that his unit would soon be in action against the enemy. Right on cue—at 1800 hours—McInnis spotted an unidentified tank about a thousand yards from his position. The lieutenant radioed headquarters to check if there were any friendly armored units in that area and was given the unhelpful reply: "What you are looking for is in front of you." This could almost qualify as one of the great misleading statements in military history. What was McInnis looking for? His inquiry was about friendly tanks. Was he looking for them? Or was he looking for enemy tanks? And how could headquarters be certain about what McInnis was looking for? In the event, the tank provided an apparent answer and saved the young lieutenant from racking his brains about the meaning of headquarters' enigmatic reply. The tank set off to the north, spraying the First Platoon positions with machine gun fire. Soon seven other tanks were seen, which also fired their machine guns at McInnis's men and opened up with their 75mm turret guns. At the distance of a kilometre and in the drizzly, misty conditions, McInnis was unable to make a positive identification of the tank types, but their fire was real enough to leave him in no doubt that he was in contact with the enemy. As a result, he ordered his platoon to open fire. In fact, his men were itching to hit back at the tanks, one of them already having been slightly wounded by their fire, and with the first shot they disabled the lead tank, leaving it pouring smoke and flames. Two or three more shots were fired, but no more certain hits were achieved. But the tanks now headed straight for McInnis's unit, firing all the time. When they were no more than four hundred yards away, one of the antitank gunners identified the tanks as Shermans and called on everyone to stop firing. But the tanks kept firing, and they kept coming. Although Sergeant Nunn courageously stood up and waved at the tanks, he could do nothing more to stop them as they closed in, and his men dived for cover. The tanks simply crashed through First Platoon's position and continued northward. McInnis's men suffered only light casualties, but a nearby unit led by Lieutenant Raney was not so lucky. As the tanks rumbled past, one of his men was killed by a direct hit from a 75mm shell, and two others were wounded. The Second Platoon of Company C, led by Lieutenant Connors, also fell foul of the tanks. At a range of just fifteen yards, one of the Shermans pumped bullets and shells into an unarmored half-track, severely wounding the driver. At such a range, it is inconceivable that the tank crews could have failed to realize that they were destroying American vehicles and killing American troops. It was easy for the soldiers of the 823rd to identify individual tanks, one of which was No. 25 from the Third Armored Division, and another that carried the name "BEBACK" on its hull. What is incredible is that nobody in the tanks could make a similar identification of the vehicles and equipment so liberally marked with national and unit insignia. So ferocious was the tank attack that at one stage, Raney,

McInnis, and several others tried to shelter behind a stone building. But as if compelled to exterminate everything in its path, one tank blew the building down with high explosive shells. At a range of just twenty feet, one of the tanks turned its main armament on Lieutenant McInnis, but after a heart-stopping moment, it did not fire and moved away. As the seven remaining Shermans moved off northward, they opened fire on another American unit—the Thirtieth Reconnaissance Troop under Lieutenant Curry—and blew a hole in one of their M–8 halftracks.

This incredible incident had lasted under half an hour, and yet it seems to encapsulate all that is inexplicable in amicide. However much one allows for poor visibility, battle fatigue, the fog of war, or any other comfortable euphemism, one is left struggling for an explanation for what happened here. The casualties and matériel losses had fortunately been relatively light, but the 823rd was, in the words of Major Lohse, "mad as hell." He described the aftermath of the engagement, by saying that the unit "took two prisoners which were its first, suffered its first fatal casualties, was shot up by its own Infantry and Armored Force and in turn shot up our own Infantry and Armored Force but under all circumstances came through their first critical engagement in fairly good shape and without too serious losses." The following day, Major Lohse carried out an investigation of the curious "battle" with the Shermans. In the first place, he needed to find out why McInnis's platoon had opened fire on the Shermans. His conclusion was that McInnis had been given information by retreating infantry and by headquarters that enemy tanks were closing on his position. The poor visibility on July 9 made identification difficult or indeed impossible at a range of a kilometer, and the 823rd had received no reports of friendly tanks in the area. Once the tanks opened fire, Lohse felt, McInnis was entitled to assume that the mystery tanks were unfriendly, and when the tanks began heading north when the main movement of all Anglo-American thrusts was at that time in a southerly direction, the 823rd could legitimately assume that they were German. On the other hand, when the tanks were within four hundred yards, the 823rd gunners were able to identify them as American and enact an immediate cease-fire. When this did not stop the tanks, several individuals tried to identify their units as American, but to no avail. Nothing seemed to stop the tanks, and Lohse was at a loss to understand the thinking of the tank crews, who had every opportunity, notably when within a few yards of the 823rd positions, to see that their opponents were American.

The tanks, of course, were those sent by Brigadier General Bohn to take Hill 91 with all possible urgency. It would be kind, if hardly complimentary, to the tank commander to suggest, as Charles Shrader does, that he "became confused." He had apparently taken a wrong turning at a crossroads, and instead of heading toward Haut-Vents, he blundered straight into the 823rd TD Battalion position. It can hardly have helped when the lead tank, presumably containing the commander, was knocked out by McInnis. Without proper guidance, the other tank crews seem to have lost their heads. Once the remaining Shermans had passed through 823rd's lines, they reversed their direction and headed back toward Hill 91 at Haut-Vents, which they reached as darkness fell. Incredibly, no sooner had the tanks actually arrived at their correct position than they were attacked and strafed by American aircraft. Apparently, these strikes had been called in earlier when the hill was occupied by enemy troops, but having been delayed by bad weather, they were eventually launched when the Germans had gone and been replaced by friendly units. This illustrated the problems for amicide of an increasingly fluid battlefield.

The problems faced by the 823rd were only the tip of a large iceberg. In the early period of the fighting in Normandy, before the breakout, Allied troops were so crowded together that there was bound to be a degree of confusion and friendly casualties. On July 10, the Thirtieth Division again came under fire from American tanks, and five days later a more serious blue-on-blue occurred north of St. Lô. As the Twenty-ninth Division began its assault on the French city it found itself under fire from units of the neighboring Thirty-fifth Division.

During the Battle of Schmidt in November 1944, there were many incidents of inaccurate American artillery fire, with consequent friendly casualties. American troops of the 112th Infantry were in retreat from their positions in Vossenack, intending to reestablish their lines farther back. Before they left, the infantry had summoned artillery fire, but to their consternation, when it came, the first four volleys of shells landed right on the new defensive line that was being set up. One shell demolished a barn in which soldiers from the First Platoon of E Company were sheltering, killing one man and seriously injuring three others. The commander of E Company, Lieutenant Melvin Barrilleaux, rushed to the command post to try to stop the guns, but before he had gone more than a few yards, another shell killed his sergeant and wounded him. Barrilleaux and the wounded men were evacuated to a rear first aid station and by the time the friendly fire was halted, the entire American position had been fatally weakened and the retreat continued.

The difficult terrain and the harsh weather conditions made artillery support a questionable commodity at this stage of the European war. Shrader cites several other examples from this period, including one on December 14, when men of the Ninth Infantry near Wahlerschied were hit by their own artillery. Later in the same week, two battalions from the 343rd and 344th Infantry endured a heavy fire from American artillery that so shattered their morale that they nearly panicked and abandoned their posts. Eventually, the American guns were only stopped by an officer running ahead to an observation post and signaling their predicament. In fairness to the American troops engaged in a confusing and bitter struggle in appalling conditions, Charles Shrader evens things up by citing examples of German friendly fire. On October 4, 1944, near Uebach, troops from the German Forty-ninth Infantry Division were hit by their own guns, while on December 28, not far from Sadzot in Belgium, the Twenty-fifth Panzer Grenadier Regiment called up artillery support and found that its own mortars were dropping shells into their own lines. The problems of a night battle and the difficulty of communicating precise references were clearly at fault here.

An unusual blue-on-blue involving armored forces occurred at Oberembt on the night of February 26–27, 1945. The American Thirtieth Infantry Division was engaged with units of two German divisions, the Ninth and the Eleventh Panzer along the Roer River. As the flat terrain provided poor cover for tanks, most of the armored operations were taking place at night, improving the possibility of concealment, but also increasing the likelihood of misidentification. The 117th Regiment had already taken the town of Oberembt, and the plan was to push on that night to seize two villages and the town of Putz. As night fell on February 26, the Third Battalion of the 117th was detailed to take the village of Kleintroisdorf, while the First Battalion simultaneously captured Kirchtroisdorf. Once these targets had been captured, the First Battalion would move through its sister units and go on to take Putz. To overcome the danger from minefields, the infantry was to be supported by American tanks from the 743rd and British flail tanks of the Lothian and Border Yeomanry, which had been specially designed to precede the infantry exposing and exploding mines with their flailing chains.

The operation started well, and by midnight the two villages had been captured and the Second Battalion had begun its advance to take Putz. But the success of the mission was compromised by an unfortunate incident involving the British and American tanks. The four British flail tanks had been ordered to support the First Battalion's attack on Kirchtroisdorf; instead, they turned left instead of right and followed the Third Battalion toward Kleintroisdorf. After traveling a few hundred yards, the platoon leader, realizing his mistake, turned his tanks around and headed back in the right direction. As they did so, they were spotted by American tanks and wrongly identified as German. The American tanks opened fire and destroyed all four of the flail tanks before they could identify themselves. As usual, human error was at the root of a blue-on-blue. A simple navigation error, coupled with poor visibility and a failure to identify friend from foe, produced a tragic and wasteful incident.

War in the Pacific

The ground war in the Pacific—either the island-hopping operations of the Americans or the jungle fighting of the British in Burma and the Australians and Americans in New Guinea—contained at least as many examples of friendly fire as the more traditional fighting in the European theater. Added to the normal ingredients for disaster, such as inexperienced troops, poor visibility, misidentification, poor coordination, difficult terrain, and a hostile environment, must be added the cultural shock brought about by a conflict between American and Japanese codes of behavior. Many green American troops regarded the Japanese as quite unlike themselves: either subhuman through their capacity to survive in the jungle on a handful of rice, or superhuman in the early years of their triumphs in Southeast Asia. Not until later in the war did either American or, for that matter, British troops come to a realistic estimate of their Japanese opponents. This contributed to the feeling of unease felt by American troops in their early engagements with the Japanese and resulted in many blue-on-blue incidents. General Eichelberger described one particular characteristic of American troops: "Excitedly firing at noises during the night was a common fault and seriously restricted the use of patrols and other important movements after dark." Indiscriminate firing—later to be a problem in Vietnam—was thus a habit that had to be addressed in an attempt to reduce friendly casualties. On February 25, 1944, a report by Colonel Horace Cushman contained the following:

Killing or wounding our own troops

> During recent operations a number of officers and enlisted men have been killed or wounded at night by our own troops who fired with the belief, or from the fear, that the Japs were infiltrating into their areas. The majority of cases reported occurred among troops bivouacked well to the rear of the front line infantry battalions. Some of the men and officers were sleeping in their jungle hammocks when shot. The majority of this "trigger happy" firing, although not restricted to troops which had not been previously in action, was among newly arrived units.
>
> Officers with battle experience in this theater are of the opinion that this condition is contributed to by the overemphasis placed on the ability of the Jap to infiltrate into our rear areas and by the oftenheard statement "stay in your slit

trenches after dark, assume that everything that moves is a Jap."[26]

Fear of the unknown or, for many green American soldiers in 1942 or 1943, the unknowable, in the shape of the Japanese, was at the root of these early friendly fire incidents. What was rarely understood, of course, was that the Japanese were a basically urban people and most of their troops were no more accustomed to living and fighting in a jungle than the Americans. In such a case, the Americans were victims of their own cultural stereotypes. It was not until the Americans and the British came to terms with their own prejudices about Asian people in general that they were able to encounter them on equal terms. It is noticeable how this problem had still not been overcome by the 1960s and America's long involvement in Vietnam. Even by then, American troops had not learned that the enemy should neither be underrated nor overrated. A soldier should fight the enemy that faces him, not the one he imagines, otherwise he is a victim of his own fears. This "painting of pictures" has hindered the performance of European and American troops in their conflicts with Japanese and Asiatic troops in the twentieth century.

One of the most extraordinary instances of friendly fire—and one closely linked to the problems of "painting a picture"—occurred on August 15, 1943, when large numbers of American and Canadian troops were landed in the Aleutians, on the island of Kiska. So severe were weather conditions in the area, with frequent storms, persistent fog, and heavy snow, that for weeks at a time it had been impossible even to locate the island, and through a serious intelligence failure, the American commanders expected to face fanatical resistance by Japanese troops, even though there were, in fact, none whatsoever on the island. The Allied troops selected for this operation had never seen combat before and had obviously been building up to expect the struggle of a lifetime against enemies larger than life, tenacious, resilient, and fanatical. On August 15, a huge armada of ships, including two battleships and five cruisers, converged on Kiska. The island had been subjected to more than a fortnight of bombing—when the aircraft could even locate the island—and by nightfall thirty-five thousand troops had been landed. By the end of a day of sustained fighting, twenty-eight American servicemen were dead and a further fifty seriously wounded. Most of the casualties had fallen victim to their own comrades during gunfights taking place in a thick mist. Upon landing, the American troops fanned out in the heavy fog, expecting to face enemy resistance at every turn. The American columns frequently exchanged fire, and when night fell, the inexperienced soldiers lost even the confidence that daylight had given them. A prey to their own fears and those of their officers, they fired at every walking shadow. As one survivor, Lieutenant Murphy, later wrote: "The troops were shooting at anything that moved." One American soldier, convinced he was attacking a Japanese unit, had to be deliberately shot down by his comrades as he insisted on charging and flinging grenades as he ran, even though they shouted at him to stop in English. The Kiska fiasco was written off by the commanders as good experience. Casualties were unfortunate, but it had been a good exercise with live ammunition. As Admiral Kincaid, the invasion commander, later wrote: "We had no way of anticipating our men would shoot each other in the fog." As we have seen before, this is not a fair comment. Locate green troops in an area of expected enemy activity, add fog to the equation, and one gets a situation ripe for blue-on-blue. Admiral Kincaid should have known better than to blame it all on the men.

The American soldiers found themselves completely disoriented in the islands of the Pacific. As one corporal wrote: "Get used to noises at night. This jungle is not still at night. The land crabs and lizards make a hell of a noise, rustling on leaves. And there is a bird here that sounds like a man banging two blocks of wood together. There is another bird that makes a noise like a dog barking." The result was that many GIs found it difficult to sleep and were constantly alert for the slightest sound. They fired indiscriminately and caused many friendly casualties. One war correspondent wrote about Japanese scare tactics:

> The constant roar of artillery and mortars made it impossible to sleep at night. In addition, the Japs used night harrassing tactics. Most of the night fighting was done with knives and

> machetes. Muzzle blast from rifle fire at night gave away positions and locations of troops. During the night men rested in foxholes three or four feet deep. Usually there were four men in a foxhole, sometimes less. Japs sneaked in pairs towards the foxholes. One would often jump into the middle of our men and try to stab them. The others stood by to see the outcome. Sometimes the Japs would jump in the foxhole and then jump out quickly hoping our troops would become excited and stab each other.[27]

On New Georgia, there was an outbreak of friendly fire among green troops. Terrified by Japanese psychological tactics, American soldiers sprayed bullets about and even threw grenades, killing a number of their comrades. On Hollandia, units of the U.S. Twenty-fourth Division lost all control and fell victim to their fears. According to General Eichelberger, the men who had been ordered to guard his headquarters:

> . . . carried on a terrific war. Tracer bullets from all directions made fireworks in the camp. Automatic rifles were fired and grenades were thrown. Troops on the inside and on the outside thought they were being attacked by the Japanese . . . the battle of Brinkman's Plantation was . . . a battle among Americans. A master-sergeant was killed and a number of troops were wounded. It was a disgraceful exhibition.[28]

General Omar Bradley, when questioned about the main shortcomings of inexperienced troops, listed "reliance on rumor and exaggerated reports" as one of the most debilitating factors. This weakness was clearly in evidence during a large number of the Pacific friendly fire incidents. It was only necessary for the word to spread that there were Japanese troops on the loose for green troops to begin spraying bullets about. It was not so much the Japanese soldiers themselves who inspired the fear, it was the idea of them, the strangeness and the fierceness of their behavior, the fanaticism and cruelty with which they fought, their willingness to give their lives and fight to the end. In open combat, the American soldiers were at least the equals of the Japanese, but in the jungle environment of so many of the Pacific battlegrounds, the American soldier felt isolated and alienated. As he fired his bullets indiscriminately into the tropical vegetation, it was not just the Japanese soldier that he was trying to keep at bay, but the primal fear he felt in himself of darkness and the unknown. American military training failed to help their men to adjust to jungle fighting, in the Pacific, in New Guinea and later in Vietnam. A British sergeant, later reflecting on American training methods, made the following valid observations:

> It appeared that the American infantry men were not trained in "battle noises." They seemed to drop to the ground and fire wherever shots were heard close by . . . It was purely a matter of lack of experience. They were shouting at each other and firing at nothing.[29]

A common feature of friendly fire incidents in the Pacific War was poor coordination between different units, notably between army units and the marines. During the invasion of the Gilbert Islands in November 1943, the planners of the Twenty-seventh Division devised a system for occupying the island of Makin that almost guaranteed friendly casualties. The island was divided virtually in half by a deep and wide trench, known as the West Tank Barrier. To overcome this obstacle, the plan was for the First and Second Battalions of the 165th Infantry Regiment to coordinate an attack from both east and west simultaneously. To avoid friendly casualties, the two battalions would stay in radio contact throughout and signal to each other with colored smoke flares. It must have seemed a foolproof system to the planners aboard some U.S. naval unit in the Pacific. But to the battalion commanders on Makin—Lieutenant Colonel Gerard Kelley of the First and Lieutenant Colonel John McDonough of the Second—it soon became a nightmare. In the first place, each failed to establish radio contact with the other. Next, the First Battalion soon found itself pinned down by what they assumed was friendly fire from their sister battalion. In spite of every effort to get the Second Battalion to stop firing, Kelley was ordered to get his men moving forward to meet up with McDonough. While the Americans were making things harder for themselves, Japanese snipers were adding to the confusion by picking off stragglers. So panicky did the American soldiers become that they kept firing throughout the night of November 20–21, inflicting more friendly casualties and wasting huge quantities of ammunition, without ever having much idea where the Japanese defenders were. Dawn on November 21 only brought more absurd incidents. Landing craft approaching Yellow Beach machine-gunned two empty hulks, which had already been rendered uninhabitable by air attacks. But, convinced that they harbored Japanese troops, first the LCIs strafed them, then further air strikes were called in to attack them, and finally tanks were brought onto the beach to bombard them with their main armament. Apart from being a waste of ammunition, this mad attack on phantom enemies resulted in many friendly casualties, as bullets and shells went over or through the hulks and wounded American troops in the jungle behind. The problem was that the troops were mainly inexperienced National Guardsmen, in action for the first time, who allowed themselves to be pinned down by even small pockets of Japanese defenders. At night, they were demoralized by Japanese scare tactics, including simple techniques like uttering threats in English, throwing firecrackers, and keeping up a regular sniper fire. The American troops of the Twenty-seventh Division were thoroughly rattled and kept firing indiscriminately at unseen targets, exposing themselves to counterfire. As one soldier wrote:

> Smoking out the snipers that were in the trees was the worst part of it. We couldn't spot them even with glasses and it made our advance very slow. When we moved forward it was as a skirmish line, with each man being covered as he rushed from cover to cover. That meant that every man spent a large part of his time on the ground. If one of our men began to fire rapidly into a tree or ground location, we knew that we had spotted a sniper, and those who could see took up the fire. When we saw no enemy we fired occasional shots into trees that looked likely.[30]

Some of the Americans lost their nerve completely, and one ran along the beach shouting: "There's a hundred and fifty Japs in the trees." This only added to the hysteria as the men of the Twenty-seventh peppered the trees and each other with wild fire. It took nearly four days to complete the occupation of the island of Makin, during which time Japanese submarines exacted a heavy toll of American support ships that should have been able to move away from the island in the first twenty-four hours.

The coordination of artillery fire was sadly amiss during the Buna campaign in Papua, which Charles Shrader refers to as a "Leavenworth Nightmare." Cooperation with Australian forces in the operation were stretched when a mortar shell from an American battery fell on a command post occupied by Captain Jack Blamey, nephew of the famous Australian First World War commander, General Sir Thomas Blamey, killing him and another soldier and wounding six more.

A further dimension to friendly fire in the Pacific war was the role played by naval gunfire in supporting American ground troops. The situation for the naval gunners was fraught with difficulties, not least because they were always firing blind and frequently had to find their targets after bombing runs by aircraft. The consequent smoke and dust clouds made targeting almost impossible. On February 22, 1944, American ships were due to support a landing at Parry Island. The planes had already plastered the Japanese defenses, and much of the island was masked by smoke. As the landing craft approached the shore, they disappeared entirely into the murk and unfortunately were hit by shells fired from the American destroyer *Hailey,* resulting in thirteen deaths and forty-seven men wounded. On the same day, further five-inch shells from American destroyers hit friendly troops and destroyed a number of tanks. In view of the fact that the naval gunfire had helped in the suppression of Japanese resistance on the island, the friendly casualties were accepted as the price that has to be paid for such indirect fire in difficult conditions.

In April 1944, during the occupation of Bougainville, Company K of the all-black Twenty-fifth Regimental Combat Team, on its first patrol, encountered a Japanese machine-gun position. Coming under enemy fire for the first time, the men of Company K panicked and began firing wildly and then more carefully at each other. In spite of every effort by the company commander to stop the firing, by the time the company fled in disorder it had suffered ten men killed and twenty wounded, many of whom had fallen to the bullets of their comrades. The American operation on Saipan in June 1944 was controversial less for the numerous incidents of friendly fire than for the relief of Major General Ralph Smith, commander of the Twenty-seventh Division, by Lieutenant General Holland Smith, commander of the Fifth Amphibious Corps. The dispute revolved around the slow and uncertain performance of Ralph Smith's men, which had already been seen during the occupation of Makin. On Saipan, the Twenty-seventh was a frequent victim of its own artillery. But what made matters worse was that the Twenty-seventh's artillery also distributed its shells on neighboring marine units, causing heavy casualties. In Charles Shrader's opinion: "Although Ralph Smith's relief was ostensibly based on the slow and uncoordinated advance of his division on June 23, there can be little doubt that Holland Smith's decision was influenced by the unwarranted shelling of his marines by the Twenty-seventh Division's artillery." In a sense, therefore, Ralph Smith was himself a victim of friendly fire.

The Americans were not the only troops on Saipan guilty of friendly fire. Japanese soldiers, unwilling to accept defeat and surrender, chose suicide for themselves and killed any of their own people less eager than they to follow the military code of honor. Japanese civilians, mostly women and children, had been fleeing from the advancing Americans until, on the cliffs at the northern tip of the island, there was nowhere else to go. It is believed that as many as two-thirds of the twelve thousand civilians on the island committed suicide by drowning or leaping from the eight hundred-foot cliffs. Japanese snipers shot down any of their people who tried to surrender.

The occupation of Guam in July and August 1944 was fraught with difficulties of coordination between army and marine units, as on Makin and Saipan. The dense jungle and fanatical Japanese resistance was responsible for holding up the advance of the Seventy-seventh Infantry Division to capture Mount Barrigada. Part of the trouble was the roadblocks set up by the Japanese. As a result, a platoon from the 307th Regiment, supported by a tank, was ordered to pass down Finegayan Road to clear them. Having successfully removed two blocks, the tank opened fire on a suspected third roadblock, only to find that it was manned by men from the Third Marine Division. The marines had been warned that the army patrol was coming, but clearly the patrol had not been told to expect the marines. The army men had expected friendly forces to identify themselves with red smoke, but nobody had told the marines. The army patrol would not stop firing until the marine company commander risked his life by waving his helmet and running down the road toward them. Seven marines had been wounded in the snafu.

A few days later the marines got their revenge. A patrol from the Second Battalion of the 306th Infantry was moving along the Salisbury Road when it came under rifle fire. The suspicion was that the fire was coming from the Third Marines, and this suspicion became a certainty when the rear company of the Second Battalion got into a firefight with marines at a road junction. When a complaint was lodged at Third Marine Division headquarters, the army was told categorically that there were no marines in the area of the incident. No sooner had this reply been received by the army than the command post of the 306th Infantry Regiment came under fire from marine howitzers, while another army patrol was machine-gunned along Salisbury Road by men suspected of being marines.

It seemed that the Japanese were not needed; the Americans were content to fight themselves. Coordination between army and marine units had reached a new low point, and it seems impossible to attach an "accidental" label to all of these incidents. On August 8, units from the Seventy-seventh Infantry Division, the First of the 306th and the Third of the 307th, engaged in a costly firefight. Each had come under mortar fire from the general direction of the other and, apparently using this as justification and making no attempt to identify the enemy—who could have been Japanese, after all—they returned mortar fire. Each radioed that they were engaged with a Japanese counterattack, and then began spraying bullets in the general direction from which the mortar shells had come. As the firing increased, both battalions became increasingly certain that this could not just be friendly fire but must now involve the Japanese. Tanks, supporting the 306th Infantry, joined in and shelled the 307th's position. As if this was not enough, each battalion then called for artillery support, and a barrage was fired by the 902d Field Artillery Battalion. Eventually, the situation was brought under control, but not before both battalions had suffered significant casualties. It was another example of the effects of mass hysteria. Officers who should have known better shot first and looked for excuses afterward.

THE KOREAN WAR

Friendly fire incidents in Korea in 1950 resulted from the poor training and often low morale of an American army that was raised hastily and had to learn its profession on the battlefield rather than the parade ground. A few examples selected from the depressingly large number illustrate the difficulties a pampered peacetime force encountered in coming to terms with the harsh realities of a war situation. On July 7, 1950, the Third Battalion of the Thirty-fourth Infantry, in action with North Korean regulars near Ch'onan, was targeted by its own artillery and hit by mortar shells. Only when Major John Dunn went to the rear himself was the fire brought to a halt. Later, as the American troops were forced to retreat, men of the Twenty-first Regiment were bombed and strafed by American planes.

Later in the year, with the intervention of the Communist Chinese, matters became even more desperate. The American retreat from the Chongchon River revealed an army in disintegration. Green troops sprayed bullets in all directions, and a machine gunner, in a state of shock, turned his weapon on his own colleagues. The fact that the American troops were part of a United Nations army only served to make things more difficult. With fourteen nations committing combat troops, misidentification between allies made amicide a strong possibility. To make matters worse, there were Koreans on both sides, and this latter point resulted in a disastrous blue-on-blue on November 26, 1950. The South Korean Second Corps, fighting on the right flank of the American Second Division, was broken by Chinese attacks and forced to retreat southward. During the early afternoon, the commander of the U.S. Thirty-eighth Infantry Regiment, Colonel Peploe, witnessed a mass of disorganized Korean soldiers fleeing through his lines. It was the Third South Korean Regiment, in full retreat. But the American soldiers had not been told to expect Koreans heading south, and opened fire on them, believing them to be assault troops of the North Koreans. After some difficulty, Peploe managed to halt the American friendly fire. But at American divisional headquarters, the decision had already been taken to bolster the front weakened by the South Korean collapse with troops from the Turkish Brigade, newly arrived at the front and quite unaware of the identity of the fleeing Korean troops. No senior American commander briefed the Turks as to their task; they were simply thrown in at the deep end to sink or swim. Undeterred, the Turks—five thousand strong—immediately set about fighting the first Koreans they met. Unfortunately, they turned out to be the fleeing South Korean Third Infantry Regiment. Near the village of Wawon, the Turks came under fire and responded fiercely. Soon the American Second Division headquarters was deluged with reports from the Turks that they had just routed the enemy, inflicting many casualties and taking numerous prisoners: Unfortunately, they were the South Koreans. While the Turkish Brigade was licking its wounds and relishing its victory, it was suddenly hit by overwhelming numbers of Chinese. For two days, it hung on, but when it eventually fell back, it was found to have been so badly depleted that the Americans felt obliged to apologize officially to the Turkish Government. A further setback occurred the next day when the South Korean troops attached to Colonel Peploe's Thirty-eighth Infantry command tried to launch a counterattack against the Chinese and were brought under a crushing fire from nearby American tanks.

One of the most tragic cases of amicide on record occurred during the Korean War. While fighting along the Imjin River on April 23, 1951, the British Twenty-ninth Brigade, containing the First Battalion of the Royal Northumberland Fusiliers, the First Battalion of the Royal Ulster Rifles, and the First Battalion of the Gloucestershire Regiment, was ordered to withdraw from its position to avoid being overrun by heavy Chinese concentrations. Most of the brigade got away, but the Glosters were surrounded by a full regiment of Chinese regulars. They soon ran short of ammunition, medicines, and food, and although American planes tried to drop supplies, most of them fell into Chinese-occupied areas. Wave after wave of U.S. fighter-bombers swept in, drenching the rocky hills with napalm, but in spite of artillery and air support, the Chinese tightened their grip on the beleaguered British troops. Throughout April 23, the Glosters beat off attack after attack by waves of Chinese, but at dawn on the twenty-fourth, A Company was overrun, while B Company was reduced to just one officer and fifteen other ranks. Of the original 622 officers and men, the Glosters had been reduced to a small but tough group of men holding a perimeter of just a few hundred yards. Lieutenant Colonel J. P. Carne, commanding the battalion, called for helicopter support to evacuate his wounded, but the enemy fire was too fierce to risk bringing the choppers in. On the afternoon of April 24, several attempts were made to relieve the British, but they were driven back by overwhelming Chinese numbers. As night fell on the twenty-fourth the Chinese renewed their attacks, but all through the hours of darkness, the Glosters drove the enemy back, until the hillsides surrounding their position were covered with thousands of enemy dead. At dawn on April 25, Carne's battalion was down to fewer than three hundred men, and ammunition was so low that the British only fired when the Chinese were within fifteen yards of their lines. With the rest of the Twenty-ninth Brigade in full retreat, the Glosters were now entirely on their own, in a sea of Chinese troops. There was nothing left for the survivors but to try to break out as best they could. Colonel Carne again called in air support, and once again the U.S. dive-bombers answered the call, blasting out the Chinese from within thirty or forty yards of the British position. Yet however many Chinese fell, there was still no way out for the Glosters, and Carne called his company commanders together and offered them the choice of surrendering or trying to fight their way clear. Most chose the second option, though there was no chance of taking the numerous wounded men with them, and Carne himself volunteered to stay with the injured. With Carne, the battalion doctor, and the chaplain, stayed the Regimental Sergeant Major Hobbs, a man who had spent his whole adult life with the regiment.

Once the decision was made, the remnants of B, C and D

companies set off southward to try to break through Chinese lines. Captain Michael Harvey led the one hundred men of his group north instead, then westward, circling around the Chinese positions, before heading south to where he hoped to find friendly positions. At first, Harvey's group was lucky, wiping out small Chinese sections, but encountering no large enemy units. But their luck did not hold. Moving down a valley, they found the hills on both sides swarming with Chinese. Harvey later estimated that forty artillery pieces opened up on his grim band, and several of the Glosters were killed or wounded. Yet Harvey had made it clear from the outset that there could be no stopping for casualties; every man knew that. The survivors of his party found refuge from the Chinese fire in a foot-deep ditch that ran along much of the valley. Still men fell while the rest struggled on, their hands and feet bleeding from the jagged rocks over which they were forced to crawl. Just as the survivors felt their suffering would never end, they saw a sight that must have lifted their hearts: about five hundred yards away was a line of American Sherman tanks, stretched across the valley, which were firing at the advancing Chinese troops. Harvey and his men struggled to their feet and ran and crawled toward the tanks. The young American lieutenant in charge of the tanks had received no instructions to keep a lookout for friendly troops in the area and certainly would not have expected to see them coming toward him from the direction of the Chinese. As Harvey and his men got closer to the tanks, the lieutenant ordered his Shermans to open fire on them, with machine guns and main armaments. In seconds, at least six Glosters were cut down. It was a tragic case of misidentification. An American plane, identifying the men as British, swooped down low over the tanks, waggling its wings and trying to distract the Shermans from their prey, but the lieutenant continued firing.

Captain Harvey, lying flat on the ground to allow the tank fire to pass over his head, found a stick within reach, stuck his cap on it, and crawled forward, waving it at the tanks, but to no avail. Trapped between a pursuing enemy behind and friendly fire ahead, many of the exhausted Glosters were caught and bayoneted by the Chinese. Desperate to end this atrocious blue-on-blue, the circling U.S. plane flew low over the tanks and dropped a hastily scrawled note. The truth at last dawned on the unhappy U.S. tankers, who realized they had been machine-gunning friendly forces who had just fought their way out from the midst of overwhelming Chinese forces. The tanks stopped firing, and the wretched survivors were able to crawl into American lines. Together the tanks and what was left of the Glosters retreated down the valley to safety. Harvey would never admit to the Americans how many of his men had fallen under their fire. Harvey's thirty-eight men were the only survivors from the entire First Battalion of the Gloucestershire Regiment, known to the world ever afterward as the "Glorious Glosters" for their heroic resistance. Rather fewer people know the dismal fate of Harvey's survivors, victims of a tragic blue-on-blue; perhaps it is better that way.

A second and much more famous blue-on-blue, immortalized in print by S. L. A. Marshall and on celluloid by a famous war film, *Pork Chop Hill*, occurred during the Korean War. In the last few weeks of the fighting, in April 1953, American and Chinese forces were contesting possession of a hill, named Pork Chop Hill. Two companies—K and L—of the Thirty-first Infantry Regiment, part of the American Seventh Division, were ordered to attack and take the hill, but from opposite sides. Poor coordination between the units contributed to a classic blue-on-blue. No sooner had Company K reached the top of the hill than they came under fire from the machine guns of Company L. Believing the hill to be held by the Chinese, First Platoon of Company L could not be convinced to stop firing until they literally ran out of ammunition.

THE VIETNAM WAR

Incidents of amicide among ground troops were so frequent during the long American involvement in Vietnam that it would be tedious to attempt to do more than indicate some of the main problems experienced there. At least 5 percent of all casualties during the war were due to accidents and friendly fire. Ten thousand Americans died from noncombat causes, including amicide, and deaths from the latter category were always listed as KIA—killed in action. According to

Charles Shrader, incidents of amicide in Vietnam "were precipitated by nervousness and lack of fire discipline or by inadequate coordination."

The American military philosophy in Vietnam, as in Korea, was to employ technology as a substitute for highly trained infantry. This fallacious policy, sometimes referred to as spending bullets rather than lives, resulted in the application of aircraft, helicopter gunships and artillery to a situation that could better have been handled by the men on the ground. This application of hardware often hampered the work of the infantry, as well as inflicting unnecessary casualties on them from blue-on-blues. However, the use of firepower rather than manpower was popular with the troops and with public opinion in the United States. Although it reduced casualties from enemy fire, in many ways it revealed an underlying lack of commitment on the part of both the generals and the politicians, who feared the political effects of a high body count more than they feared the Communists.

The majority of the American troops involved were not regular soldiers and lacked a high degree of dedication or self-motivation. Most, as civilian draftees, were naturally much concerned with self-preservation, and their commitment to the military profession or the cause of saving Vietnam from the Communists was questionable in many cases. The unpopularity of the war among soldiers in Vietnam and the public in the United States only served to lower morale. Racial stereotyping, of the kind that had affected American and British soldiers in their attitude toward the Japanese in World War II, was even more marked in American attitudes toward the Vietnamese people as a whole, and this obviously contributed to a wide range of friendly fire incidents, some of a disciplinary nature, like fragging. Two brief examples from Shrader can serve to represent the kind of errors that were regular occurrences in the confused jungle fighting. In October 1966, during a patrol near Bong Son, a soldier of the Twelfth Cavalry stepped off the jungle path his unit was following to relieve himself. When he returned, he was mistaken for an enemy and shot dead by his own best friend. A second incident five years later involved a soldier who, becoming disoriented, wandered out of his unit's defensive perimeter. Finding himself lost, he returned to his lines from a different direction and was shot dead by one of his colleagues. Casualties from this cause—carelessness and panic by inexperienced troops operating in a threatening environment—make no headlines. Nor is it always possible to tell how patrol casualties were caused, whether killed by the enemy or accidentally by their friends. One can only surmise that these accidental killings must have made up a far from insignificant percentage of the total for American casualties in Vietnam.

An incident involving Australian troops offers a rarer and yet more substantial example of ground amicide from Vietnam. It emphasizes the problems of coordination between small platoon-size units and how in the difficult environment of a jungle battleground, mistakes of identification are almost inevitable. On September 19, 1971, Lieutenant Gary McKay was part of a company-size operation in thick, leafy jungle to the southeast of an area known as the Courtenay rubber plantation in Phouc Tuy Province in the very south of Vietnam. While out on patrol, McKay received a warning from company headquarters that he could expect to encounter enemy troops within a thousand yards of his position, and so he issued verbal warnings to each of his platoon before moving on. The platoon had gone no more than another 150 yards when firing broke out at the front of the column. McKay called out: "Contact front" to his signaler to inform the rest of the company, but was surprised when the signaler replied that their neighboring unit—the Tenth Platoon—had already reported that they were in contact with the enemy as well. In McKay's words: "My mouth went dry and my heart skipped a beat as it struck me what was happening. I sprinted forward screaming out for everyone else to cease firing . . ." It was a blue-on-blue. McKay's platoon had blundered into another Australian unit; itchy trigger fingers had done the rest, with inevitable results. McKay's scout had shot one of the other platoon members in the head, inflicting a shocking, certainly mortal wound. McKay remembers that as he passed by the other platoon, one of the soldiers "mouthed an obscenity at me and I turned on him and let loose with all the venom I could muster. I knew how he must have felt having lost a mate because of a mistake on my part, but I really didn't need that kind of hassle just there and then."

McKay next set out to investigate how the accident had occurred. The forward scout had apparently fired at an unidentified movement that McKay explained was exactly what his platoon had practiced to avoid. But no amount of practicing could prepare troops for the real thing, and once exposed to a situation in which troops were told they were within a thousand yards of the enemy, men were inclined to shoot first and feel guilty afterward. And, of course, the guilt felt by men who had shot comrades was something impossible to quantify or to remove permanently. McKay admitted that he felt much of the responsibility himself for having chosen the wrong track. Although he had not pulled the trigger, he could not rid himself of the feeling that he was responsible for the actions of the men in his platoon. Both McKay and the forward scout benefited from continuing on the search-and-destroy mission, although no contact was made with the enemy. It was vital that neither man was allowed too much time to think about the blue-on-blue, or else morale in the platoon would have been irrevocably reduced. McKay was pleased that there was no contact with the enemy, as he was doubtful how his platoon would respond. It was possible that they would have been so aware of the need for caution that they would have fallen into the error of being too cautious, thus reducing combat efficiency. In the event, the patrol passed off peacefully, except for the blue-on-blue.

McKay had already expressed his unease at being dependent for artillery support on neighboring American units, some of which he believed to be less than combat efficient as a result of the drugs that so many of the soldiers seemed to regard as essential to their survival in Vietnam. McKay feared that if they were to fire in support of his unit, there was every chance of a blue-on-blue. As he wrote:

> The Platoon were visiting the US 155s as they hadn't seen these huge guns before and so I took the opportunity to speak to the lieutenant in charge . . . He was a decent sort of a bloke but who had a real load on his plate.
>
> . . . I asked him quite bluntly what all the noise was the night

before. He looked at me and said, "Oh yeah, last night was our beer ration night and everyone was letting off a little steam." I commented that they must have had a truckload of beer from all the noise they were making; he said "Shit, no man, they smoke dope." I couldn't believe my ears. Here was a direct support battery of 155mm guns firing in support of us and the battery was full of pot heads! I questioned him further and asked how he controlled this sort of thing and he said that they had a weekly "shake-down" when they searched the soldiers' tents, but when the soldiers ran past his bunker on the way to muster, they would throw the dope into his bed space to avoid detection. He was quite sincere when he said that he considered he was pretty lucky since "none of my guys are onto hard stuff."

I was stunned and had to thank God that we didn't have a drug problem in our army like the Americans had. It still worried me a lot that these soldiers were likely to fire in support of us whilst we were patrolling in this area . . . I made a mental note that if and when we called for fire support in future I would try not to use that American battery.[31]

The thick foliage in the Vietnamese jungle accounted for a number of blue-on-blue incidents. In his recent autobiography, General Norman Schwarzkopf describes the cause of the incident that gave rise to C. D. B. Bryan's book, *Friendly Fire*:

On the night of February 18, 1970, C Company had dug in on a jungle hilltop. They'd made a routine request for our artillery to zero in on the trails near their position in case the Vietcong attacked during the night. One of the test rounds detonated directly above them, spraying the men with shrapnel and killing Michael Mullen and another soldier. A subsequent investigation concluded that a lieutenant at the artillery fire direction centre, in calculating trajectories, had forgotten to take into account the vegetation on the hilltop. The round had been meant to sail over C Company; instead it had hit a tree and exploded.[32]

Gary McKay describes a similar incident, involving shells fired by the Australian artillery:

> The forward observer started adjusting artillery fire as close as we could get it and the plan was that we would keep the artillery falling just in front of the assault to keep any enemy in depth to our assault line from interfering with our attack. This proved more difficult than anyone had imagined; as the trees were so tall they were catching the odd shell, and we took a few minor casualties from our own artillery fire. The wounds were not too bad—mainly shrapnel wounds to the backs of the legs.[33]

The risk of shells being deflected by tree branches is further demonstrated in two more incidents recorded by Charles Shrader. In the first, U.S. Special Forces operating on the Cambodian border in April 1968 lost four men killed and fifteen wounded when an eight-inch shell exploded in the treetops, directly over their position. On May 10, 1968, four men of the 327th Infantry Regiment were killed when a shell with a delayed action fuse hit a tree and was deflected into their position.

The computations made by artillerymen are sometimes prone to error, and this has contributed to inaccurate or short firing, with consequent friendly casualties in Vietnam and elsewhere. In addition, the miscounting of powder charges by gun crews can be fatal. Shrader cites several examples of this problem. In 1968, an inexperienced member of a gun crew selected different lots of powder for individual shells during a fire mission. The result was that the shells showed great inconsistency, falling short or well over the target and landing among friendly troops. In 1970, at a fire support base near Hue, a howitzer battery made the same mistake, resulting in a shell hitting a company from the 101st Airborne, killing a man and wounding others. But the worst case, according to Shrader, was an incident in late 1967 in which a gun crew error in handling charges resulted in charge 7 being used rather than the correct charge 4. When the guns were fired, shells landed in an American base camp, causing thirty-eight casualties. The unit that had been hit promptly responded with counter-battery fire and hit the perpetrators of the error, killing twelve men and wounding forty more. It was an object lesson in the dangers of indirect fire. Human error had created a chain reaction that lasted twenty-five minutes and cost over ninety casualties.

The confusion of the fighting in Vietnam was at the root of many blue-on-blue incidents, as Eric M. Bergerud recalled so memorably in his book, *Red Thunder, Tropic Lightning,* describing the experiences of the men serving in the Twenty-fifth Combat Division. In the first, an unconsidered action by a tank crew had serious results:

> A fight ensued, with shots right and left. Although I didn't realize it until the next day, while the shooting was going on, there was an American tank nearby. He heard the action and decided that he wanted to get into it. He fired two fléchette rounds. Unfortunately, one of the rounds went right into my platoon, and eleven men were wounded, a couple very seriously.[34]

Misidentification of friend and foe is not only a problem for artillery, miles behind the front, or aircraft operating above the clouds. It is at its most deadly when it applies to small units or even individuals. In this instance, Dan Vandenberg describes the sort of incident that must have occurred on countless occasions not just in Vietnam or the jungle fighting of World War II, but in house-to-house operations in the European theater or night patrols in World War I. What is apparent is the ever-present fear felt by the individual soldier and the ease with which mistakes can be made. Just a split second to decide whether to fire or not, and on that moment's thought your life may depend. It is hardly surprising that so many soldiers have pulled the trigger and lived, only to regret the error.

> Marsh and I were walking parallel to each other as we entered the woods. You're all pretty jittery because you know Charlie is in there someplace, and instead of staying 10 to 15 yards apart, we started coming closer together without knowing it because you can't really see through the thick terrain. All of a sudden, we came to a break in the woods, and out of the corner of my eye, I saw something. I turned and started to pull the trigger, and I was looking right at Marsh, and he was pulling his trigger. If a fly would have sneezed, that's all it would have taken for each of us to have ripped off a magazine at each other. That's

> when I saw the ultimate look of fear. When I looked at his eyes, all I saw was white. I imagine my face looked the same.[35]

And with the error came the guilt—the knowledge that you had been responsible for killing one of your comrades. As we saw in the case of the Australian officer, Lieutenant McKay, the guilt can so affect the morale not just of an individual but of a whole unit, that they become incapable of combat:

> Track 13 was in flames: They had taken a pretty heavy hit from an RPG . . . One of the guys from 3rd Squad came walking on down towards us, walking away from the burning APC . . . I leaned out of my turret and asked if anybody up there was hurt. He just looked up and said, "Dave's dead." Dave was probably my best friend I had over there, the .50 gunner on Track 13. He was killed instantly, a direct hit. I was stunned and my mind locked up on me, and I didn't want to believe it . . . So we stayed in position for maybe five minutes until some of the other tracks backed up and turned round and we figured a way to get the hell out of there . . . They carried this guy right by my APC on a stretcher. To make it worse, we were told a little later, that he had been killed by a .50 calibre round from one of our own machine guns. Either I killed him or the APC behind me did. They were the only two .50s that were in the action. I like to think it wasn't me because of the direction I was firing in relation to the firefight, but I'll never know.[36]

In the confusion of battle, accidents occur. They are tragic, but who can take responsibility for chaos? Here an American NCO is hit by an artillery round from a friendly gun:

> The VC broke through the lines. They were shooting behind you: they were shooting in front of you. When the artillery guys lowered their guns, they were firing these bee-hive rounds point blank at the charging Vietnamese. They were opening the breech, sighting down the bore, slamming in a bee-hive round and just firing. Our sergeant ran in front of an artillery piece, and the bee-hive round cut him to pieces.[37]

And as in all other wars, noncombat accidents accounted for a high proportion of casualties in Vietnam. Lieutenant Colonel Carl Nielson wrote of one:

> At the end of that day, we had a terrible accident where a box of claymore mines went off in an armored personnel carrier and killed 5 soldiers. We had fought all night and all day, had swept the battlefield, had suffered 5 or so men killed and about 10 men wounded, and had inflicted 400 casualties, and then, right at the very end, 5 guys were killed through the carelessness of somebody who had left a set of detonators in with the claymore mines. I guess that is kind of the story of Vietnam combat.[38]

THE FALKLANDS WAR

During the Falklands War, there were a number of blue-on-blue incidents, the details of which have only recently been made public. Even the high professionalism of the British troops on the islands did not prevent at least two tragic and costly mistakes. In fact, in the case of the death of "Kiwi" Hunt, an SBS (Special Boat Squadron) NCO, it was an excess of professionalism rather than its lack that cost him his life. Hugh McManners relates the incident in his recent book, *The Scars of War.* Not long after the British landing on East Falkland, at San Carlos, G squadron of the SAS (Special Air Squadron) ambushed an SBS patrol of four men led by Sergeant Hunt. The elite SBS troops had blundered into the equally elite SAS operational area, and the SAS men were following every movement of Hunt's patrol through night vision equipment. The senior officer commanding G Squadron had placed himself alongside his machine gunner, and the two men were trying to decide whether the approaching figures were friendly troops or not. The machine gunner, convinced that no friendlies should be operating in their operational area, wanted to fire first and ask questions afterward, but the officer held him in check, still anxious over the identity of the men approaching. When Hunt's men were just thirty feet away, they were challenged by the SAS officer. Hunt apparently stood stock-still and held his weapon away from his body at arm's length, as did the two men directly behind him. But the man at the back, possibly not hearing the challenge, tried to get away in the darkness. This threat to the ambush overrode the officer's hold on his machine gunner, and the man immediately opened fire, killing Hunt instantly. It was a tragic mistake, but for men operating at the peak of their profession, it was not allowed to become a matter of deep regret. The SAS reaction was that Hunt had made a mistake and had paid for it with his life. In the hard world of these special forces, Hunt had allowed himself to be ambushed and had suffered the inevitable consequences. The rivalry between the SAS and the SBS undoubtedly played its part in absorbing the tragedy of this useless loss of life, and neither squadron wanted to be seen to be giving quarter or even asking for it. As McManners observes: "Violence is the SAS trade-mark."

A second and potentially more serious blue-on-blue occurred when A and C companies of the Third Battalion of the Parachute Regiment fought a gun battle in the hills north of San Carlos Water, on East Falkland. Late on May 21, 1982, two British Gazelle helicopters were shot down in the area by Argentine soldiers, and the two companies of the paras were hunting these enemy troops when they encountered each other. A Company saw about thirty members of C Company coming around a headland and assumed that they had located the Argentine forces; unfortunately, C Company reached the same conclusion when they sighted their fellow paras. Each company commander contacted battalion headquarters, asking for a fire mission on the other. At this point, of course, the problem ought to have been recognized by the artillerymen, who should have noticed that the friendly grid references were exactly the same as the supposed Argentine ones. But as usual in blue-on-blues, human error intervened and made a tragedy out of a crisis. The grid references relayed to headquarters by A Company's commander were wrong, placing his men apparently a thousand yards away from where they actually were. Alarm bells therefore did not ring at battalion headquarters, and the gunners duly followed instructions to open fire, assuming that there were two rather than one enemy sighting. The firefight began. One of the participants, Sergeant McCullum of A Company, reported what happened next:

> So there we were, with a strong section stuck on a forward slope with three machine guns, and a whole platoon of C Company lined up with at least five machine guns, ready to shoot at them. So you carry on, you go for it! And in the fog of war it was a classic blue-on-blue.[39]

For both sides it was their first taste of battle. Hyped up and raring to go, as these two companies of tough young men undoubtedly were, they had no second thoughts, and began firing. It was what they had been trained to do, and few of them would have had any time to worry about whether the men they were firing at were really the enemy. In addition to rifle and machine-gun fire, the battalion mortars—which both young officers had called up—now began bombarding the battlefield. If A Company commander had had any doubts, he would have found it difficult to resolve them, as within seconds his radio was shot to pieces. Soon the superior firepower of C Company was driving the patrol from A Company, with nine men already seriously wounded, to find cover. But the rest of A Company, following the best military principles and marching to the sound of the guns, soon came up to the rescue. From the area of Findlays Rocks, two Blues and Royals Scimitar armored cars appeared and, in response to emergency flares from their beleaguered comrades, began firing twelve rounds at C Company. No sooner did it seem that A Company's patrol had been saved than the battalion artillery, which C Company had called up, began landing over forty shells in and around the desperate survivors.

It was a scene of chaos. On top of the slope, McCullum and his comrades were sitting ducks for the artillery, while lower down the slope, they would be machine-gunned by the rest of C Company. It was an unenviable situation. By now two more men had suffered severe head injuries, but when a Sea King helicopter came in to evacuate them, it crashed while trying to land too quickly. A second helicopter was called in, this time carrying the commanding officer of the Third Battalion of the Parachute Regiment, who had at last deduced that the battle that was raging was in fact a blue-on-blue. Orders swiftly circulated to stop all firing. There were no Argentine troops in the area, only friendly forces.

Incredibly, in view of the ferocity of the firefight there were no fatalities, though eleven men were seriously wounded. The pride of these professional soldiers had been deeply hurt, not only because they had fought a blue-on-blue but because nobody had been killed, and this seemed to reflect on their marksmanship. Sergeant French of A Company was angry because the disorder of that night challenged his profound belief in discipline and coordinated activities.

> I get more angry now thinking back on the blue-on-blue than I did at the time, when I thought of it as part of the fog of war. It need never have happened if certain individuals (not in the patrol or in the defending company) had been doing his job—however who am I to criticize?
>
> Co-ordination of defensive positions is vital. You learn that it is far more important in war than it has ever been on exercise. The effect of the tragedy was to make every man very much more careful.[40]

Human errors have contributed to ground amicide throughout history. In the words of Charles Shrader: "Sometimes incidents resulted from human failures as simple as the inadvertent pushing of a button at the wrong time, the transposition of numbers, or a mistake in arithmetic. On other occasions, the human failure was more complex in its origins, and commonly the fear and confusion so prevalent on the battlefield played a major role."

3

Blue-on-Blue in Air Warfare

The introduction of an aerial aspect to warfare is an entirely twentieth-century phenomenon. Nevertheless, though they are recent in origin, it has not taken planes long to so revolutionize warfare that command of the air has become a prerequisite of victory in modern battles. However, in terms of its threat to friendly as well as enemy troops, the airplane has introduced a new and terrible dimension to the concept of amicide. Unlike ground weapons, it seems that technological developments in air warfare have moved far beyond the capacities of human operators to control. Charles Shrader lists no fewer than ninety-nine separate cases, taken almost entirely from the records of the United States, of aircraft engaging friendly forces. Naturally, Shrader's figure reflects only the tip of an enormous iceberg. If one were to study the records of the RAF, the Luftwaffe, or the Russian air force in World War II with the same intensity that Shrader brought to his work on American sources, one would expect to be confronted by an equally large figure. In simple terms, aircraft rule modern warfare and can inflict terrible damage on friend or foe from a height and a speed that is beyond the capacity of the ground soldier to resist. The tragedies of blue-on-blue incidents in the Gulf War, notably that involving the British Fusiliers in their Warrior APCs, illustrate quite starkly how swiftly death can come from the skies. In the case of amicide, air warfare is the growth area.

Shrader has likened the effects of air amicide to those of artillery errors in previous wars. Yet, with the capacity of modern air to ground missiles to penetrate even the toughest armored vehicles, air attacks are far more severe and much more destructive. The effects of the saturation bombing on elements of the U.S. Thirtieth Division during Operation Cobra in Normandy in 1944 reveal almost total disorientation and collapse of morale among the survivors. Ground troops do not resent artillery shorts as much as they do short bombing from passing aircraft. The way that American troops renamed their airmen the "American Luftwaffe" and shot at them on sight is a clear indication of the views of the infantry.

Since 1945, technology has made far greater strides in the field of air warfare than on the ground or at sea, with planes having greater speed, greater range, and far more terrible weapons of destruction. Each of these improvements carries with it dangers for friendly forces. Clearly, the control of aircraft operating ever farther from their bases, flying faster so that decisions need to be made in split seconds, and carrying even more lethal smart weapons, some of them potentially nuclear, makes the likelihood of errors in targeting very real. In conflicts involving fast-moving ground action, the problems of ground support revolve around the difficulty of accurately identifying targets. Failures of this kind by pilots have contributed a heavy toll of friendly casualties. Desert fighting—with its notable problems of dust and sand—contributed to heavy friendly losses in the Gulf War, and even the marking of Coalition vehicles with painted signs and fluorescent colored panels did not provide a solution, which only advanced electronic tagging will ever really achieve. As Shrader writes: "It is too much to hope that a pilot, diving at 600 m.p.h. through smoke while taking evasive action and attempting to deliver area-type ordnance accurately, could instantaneously and correctly identify camouflaged friendly ground troops making maximum use of available cover and concealment." On the other hand, one should note the advice to pilots on bombing raids made in World War II, which was echoed by Chuck Horner in the Gulf: "If in doubt, don't . . ." Do not drop the bombs or fire the missiles. This is something easily said at briefings, with blood pressure normal and without the adrenaline coursing through the veins as is bound to occur to a pilot facing a potential enemy in a "you or him" situation. Then, with your life on the line, how easy is it to remember instructions not to fire if there is any doubt?

THE FIRST WORLD WAR

Although there were many friendly fire air incidents during World War I, few of them achieved notoriety. The slow speed of the attacking aircraft and their limited bomb capacity reduced their ability to cause serious casualties when set against the ossuary of artillery victims. Nevertheless, two curious examples were recorded at the time, the first by British General Sir Edward Spears in 1914. During the early weeks of the war, the French sought to exploit the advantage they enjoyed over the Germans in air warfare. Short of a really effective aerial bomb, they employed a curious device like a large metal container filled with tiny steel flechettes, which could be tipped onto ground troops below. The weapon was abandoned when, in error, one French pilot dropped the flechettes onto a unit of French Zouaves. The effects were surprising—indeed, they almost qualified as a form of chemical or bacterial warfare. While in their container, the flechettes were immersed in oil. However, when they were tipped out, they still retained the oil on their sharp edges and so infected the wounds they caused on the wretched Zouaves, contributing to blood poisoning. Curiously, at the very time that the French abandoned these flechettes and the British adopted heavy darts, the Germans took up the French idea and used flechettes until they had perfected a suitable hand-operated bomb.

The second notable friendly fire incident involving aircraft also concerned the French. On the morning of May 12, 1916, the French airship AT-O took off from Le Havre and proceeded along the coast, looking for German U-boats. There was a mist over the sea, and visibility was not good, so that from a height of sixteen hundred feet the airship commander, Lieutenant Saint Rémy, found himself in a quandary when he faintly detected a ship below him, traveling on the surface. In fact, it was the British submarine D–3, commanded by a Canadian officer, Lieutenant Maitland-Dougall, who, like Saint Rémy, was on an antisubmarine patrol.

From the airship, Saint Rémy at last made out the shape of a submarine below him, but he detected no signs of identification. At that moment, rockets were suddenly fired from near the stern of the submarine and passed close to the French airship, convincing its crew that it was under fire. This could mean only one thing: The submarine was German and had opened fire on them. Neither Saint Rémy nor any of his crew seemed to have given any thought to the possibility that the rockets were for identification rather than an attempt to ignite the airship. Instead, the French opened fire on the submarine with their machine gun, and the crewmen who had been firing the rockets scuttled back to the conning tower for safety. Soon the French saw the submarine begin to dive. As it did so, Saint Rémy dropped a total of six small seventy-pound bombs, two of which hit the submarine on the tower and forced it back to the surface. As they watched from above, the submarine foundered and began to sink beneath the waves, while a few small figures struggled in the sea. Saint Rémy was delighted that he had sunk a U-boat and took his airship down to just a hundred feet so he could confirm his kill. But once he neared the waves and called out to the men swimming in the sea, he discovered that they were British. Unable to rescue them, he flew off to fetch help, but by the time a ship reached the scene, the men had drowned.

It was a sad and quite unnecessary disaster, caused almost entirely by the problem of misidentification. The signals used by the British ship to identify itself had been quite unknown to the French commander, and although D–3's crew had painted a recognition symbol on a forward hatch, which corresponded with those agreed between the French and British authorities for March 1–15, 1916, the French failed to understand its significance. Maitland-Dougall had been confident to stay on the surface precisely because he had already identified the airship as French. Why the submarine would stay on the surface and allow itself to be bombed if it was really German does not seem to have occurred to Saint Rémy at any stage. And why the French regarded the signal rockets as offensive weapons we can only surmise. Obviously, flying in an airship was a hazardous occupation, and fire was an ever-present fear. Thus the rockets may have posed a greater threat to the

French airship than the British crew can have understood. However, once the airship opened fire with its machine gun, the submarine had no alternative but to dive to escape the strafing, which appears to have confirmed Saint Rémy in his suspicions that the ship was a U-boat. If the outcome had been less tragic, the whole incident could be dismissed as a farce, a kind of comedy of errors. But in wartime, few things are merely funny, and Saint Rémy's errors played a major part in the loss of a British submarine and its entire crew.

The novelty of seeing aircraft employed in reconnaissance operations never quite wore off for many of the poorly educated peasant-soldiers of the tsarist armies. There are several reports of Russian aircraft being shot down by Russian infantrymen, who wrongly identified them as German. Their reasoning for shooting the planes down is interesting. When questioned by an officer, one group of soldiers claimed they had shot at the aircraft because they believed that it must be German rather than Russian, because no Russian was clever enough to design such a thing.

During the great German offensive of March 1918, when the British Fifth Army was shattered by Ludendorff's decisive stroke, the Royal Flying Corps played an important part in helping to support the ground troops. Unfortunately, as French troops were brought in to bolster the British lines, the British airmen frequently mistook the powder-blue uniforms of the French for the *feldgrau* of the Germans, causing them to attack the French with bombs and machine guns. Where they were spotting for the artillery, British pilots sometimes reported that German troops were operating behind British lines and also erroneously called down artillery fire on French troop concentrations.

THE SECOND WORLD WAR

The vast majority of examples of air amicide occurred during the six years of World War II. On fronts throughout the world, aircraft played a vital role in all aspects of the fighting, both on land and at sea, and they contributed as well to the swelling numbers of friendly casualties. Both in strategic bombing and in tactical ground support, pilots and bombardiers made mistakes that cost thousands of lives. Yet even in the early days of the war, before the scale and intensity of operations placed unbearable pressures on pilots, relatively simple errors reflected a carelessness among the young men who took to the air. On September 6, 1939, Britain's secret weapon—its newly installed radar—gave a false reading and appeared to indicate that a massive German air attack was under way. Everyone was prey to panic and the fear of the all-powerful bomber. As a result, fighters were scrambled and a tragic dogfight ensued in which two Hawker Hurricanes were shot down by British Spitfires, with the loss of one of the pilots. As if this was not bad enough, ground antiaircraft batteries joined in the battle and shot down one of the Spitfires. Clearly, at this early stage, with nerve ends showing, there were always likely to be mistakes. And again, as in almost every example of friendly fire in this book, the fundamental problem was a human one. In January 1941, during the British campaign to drive the Italians from Eritrea, the RAF strafed General Savory's Eleventh Indian Brigade, though fortunately casualties were light. But later the same year, in November, during the Crusader battles in North Africa, the First Essex Regiment suffered forty casualties when their ground-support aircraft dropped bombs among them instead of ahead of them, on the escarpment they were attacking.

Between 1939 and 1941, this problem of "early nerves" and poor identification systems was by no means a specifically British failing. The German Luftwaffe was responsible for a series of blue-on-blue incidents during the campaign against France in May 1940. Germany's blitzkrieg campaigns in Poland and the Low Countries had been dominated by the Luftwaffe, which provided close ground support for German tanks and armored vehicles. However, this close cooperation between ground and air forces was always a finely judged line, and things could and did go wrong, as General von Mellenthin experienced in Poland. He remembered how on one occasion a low-flying aircraft passed over his headquarters, whereupon his flak gunners—without attempting to identify the intruder—opened fire with all their guns. An air liaison officer rushed about trying to stop the gunners, telling them it was a German plane—an old Stork—but the gunners were so excited that they took no notice. By an incredible stroke of luck, the old plane was not hit and landed unharmed. As the gunners—now chastened by seeing the plane's identification marks—gathered around to see if the pilot was all right, out stepped the Luftwaffe general who was responsible for close-air support. As von Mellenthin observed: "He failed to appreciate the joke."

The Ju87 Stuka dive-bomber was one of the most feared weapons of the German blitzkrieg in Poland and France. One German commander was to find out what it was like to experience a Stuka attack from the point of view of the victim. On May 14, 1940, a group of Stukas mistakenly attacked Second Panzer Brigade at Querrieu, near Amiens. Their mistake was quickly brought home to them by the brigade's commander, General Heinz Guderian, who ordered his flak gunners to open fire on the German planes. Only days before, Stukas had seriously damaged a column of German tanks near Cheméry, narrowly missing Field Marshal von Rundstedt, commander of Army Group A, and Guderian was taking no chances. As he said: "It was perhaps an unfriendly action on our part, but our flak opened fire and brought down one of the careless machines." The two crewmen escaped by parachute and landed in the midst of the German infantry, who not surprisingly told the airmen what they thought of them. Guderian called off his enraged men, and having himself torn the flyers off a strip, "fortified the two young men with a glass of champagne." Guderian could afford to be magnanimous. The Stukas had done no harm on this occasion, and their work elsewhere was winning the war for Germany.

While the British and German airmen made their early blue-on-blues and learned from their mistakes, the Americans came fresh to the battlefields of North Africa in 1942 and tended to replicate the mistakes that the other combatants had made earlier in the war. One was in the field of aircraft accidents. Astonishingly in a single year—1943—2,264 American pilots and 3,339 aircrew were killed in a total of 16,128 accidents while training in the United States. So often were B–26s destroyed that a popular saying at the time at one Florida airbase went: "One a day at Tampa Bay." In case it is assumed that the Americans were alone in suffering this dreadful accidental attrition, according to German records, 45 percent of all German planes lost between 1941 and 1944 were victims of noncombat destruction. So commonly was accidental loss responsible for aircraft destruction and aircrew casualties that it threatened Germany's capacity to continue with the war.

Charles Shrader lists a large number of amicide incidents involving American aircrew in North Africa, from November 1942. Near Medjez-el-Bab in Tunisia in the last week of November 1942, a company of the American 701st Tank Destroyer Battalion, attached to the British Eleventh Brigade, was attacked by American P–38 Lightnings and lost virtually all its vehicles to air attack. After two weeks of intense effort, the company engineers had managed to repair some of the tank destroyers and get them back into action only for another wave of P–38s to attack them, strafing the vehicles and killing three men and seriously wounding two others. It is hardly surprising to find that in the unit's campaign diary, the author had some hard words for this "inexcusable" example of mistaken identity. Eight weeks later, units of the same 701st were bombed by American B–25s near Station de Sened, in Algeria. During the Battle of the Kasserine Pass, a flight of American B–17s lost their bearings, and instead of bombing German troop concentrations in the pass, they destroyed an Arab village more than one hundred miles from the battle area.

There were also examples of the ground troops striking back, wrongly identifying aircraft types and shooting down friendly planes. During the crisis of February 1943, following the American defeats at Sidi Bou Zid and the Kasserine Pass, it is hardly surprising that the antiaircraft gunners were feeling edgy. On February 21, 1943, in rain and fog, aircraft from Twelfth Air Support Command were helping ground troops of First Armored Division hold back an Axis advance toward Thala and Tebessa. Poor training in aircraft identification played a major part in the AA gunners wrecking five U.S. planes and then, the following day, shooting down a further five P–38s, in spite of a detailed warning to the gunners that low-flying friendly aircraft would be supporting ground troops. The gunners were also reminded that the enemy usually used yellow or white paint on the noses of their planes, not the black or brown of the American ones. To aid identification, the P–38s even tried rocking their wings as they flew over friendly positions, but all to no avail. Shrader is scathing about these incidents, which he puts down less to identification problems than to poor training. In view of the near hysteria that was sweeping through American troops in Tunisia at this time, after their first exposure to veteran German units, it is as well to judge the gunners with appropriate indulgence. In an attempt to prevent further losses, American ground troops were forbidden to fire at aircraft until they had themselves come under attack. The view that these inexperienced American gunners were "trigger-happy" received further support once Allied troops had left Africa and moved across the Mediterranean to Italy. Yet the enemy was hardly faultless in this regard. The Italian commander in chief in North Africa in 1940, the famous flyer Marshal Italo Balbo, was shot down and killed by his own antiaircraft gunners on returning to his headquarters at Tobruk after inspecting the Egyptian front. Flying a clearly identifiable Savoia-Marchetti S79 bomber, Balbo fell victim to the nerves of his edgy gunners, who had just received news—false as it turned out—of a mass RAF attack on Tobruk. How Balbo's single Italian bomber could have been mistaken for a "massed" anything is difficult to understand. Nevertheless, the gunners fired and achieved a friendly kill.

The Invasion of Sicily

The Anglo-American invasion of Sicily in July 1943 witnessed some of the most shocking examples of friendly fire ever recorded in military history. Mass hysteria seemed to be the problem on several occasions. Allied planners decided that the airborne attack on the island should involve four distinct and separate airborne strikes, all to be conducted by night, even though many of the American paratroopers had no night jump experience. The first assault—Operation Ladbroke—would be carried out by the British First Air Landing Brigade, led by Brigadier Hicks, which would land in Sicily from 144 Waco and Horsa gliders, towed almost entirely by American C–47s. Once landed, their task was to secure the strategic

Ponte Grande Bridge and hold it until British ground troops could move out of their beachhead. The second stage of the assault involved the airborne troops of Colonel James Gavin's 505th Regiment, who would be carried over Sicily by 266 C–47s and dropped on four zones north of Gela. Their job was to prevent any German counterattacks developing against General Patton's Seventh Army on its beachhead. The next night, more C–47s would bring in Colonel Reuben Tucker's 504th Regiment to reinforce Gavin's men at Gela. Tucker would be accompanied on the flight by assistant divisional commander "Bull" Keenens, while Matthew Ridgway, commander of the Eighty-second Airborne Division, considered too old to parachute in to take overall command of the operation, landed by sea. The final part of the operation had been allocated to the most experienced part of the Allied airborne troops, the British First Parachute Brigade, led by Brigadier Lathbury, whose job was to capture the vital Primasole Bridge.

Matthew Ridgway had become uneasy about the danger of friendly fire even before the operation began. He knew that his airborne troops would be crossing the path of the Allied invasion ships and was anxious to ensure that the ships would not open fire on the C–47s by mistake as they flew over. Quite reasonably, he addressed his views to Admiral Cunningham, the British commander of the naval operations, but found that the navy was not prepared to guarantee the safety of his force. Having fought through three years of intense naval warfare in the Mediterranean, and having suffered grievous losses from German aircraft, Cunningham was not prepared to risk allowing friendly aircraft a free flight close to his ships. Ridgway was shocked at this attitude, commenting: "We were informed that the Navy would not give assurance that fire would not be delivered upon aircraft approaching within range of the vessels at night." This should have been enough to persuade the planners to rethink the flight paths but instead they refused to change those already decided. Ridgway was even more worried to learn that his division's second strike, by Reuben Tucker's 504th Regiment, would fly not only over the invasion ships but along thirty-five miles of the invasion beaches, a sitting target for Allied guns below. All he could do was to press the navy to give him the guarantee he wanted. He warned: 'Unless satisfactory assurances are obtained from the Navy, I would recommend against the dispatch of airborne troop movements." Ridgway's courageous stand seemed to have had some effect at last when General Patton's Seventh Army informed him that the navy would withhold fire provided that the airborne troops flew along certain prearranged flight paths and came no closer than a distance of seven miles from the ships. Ridgway was relieved to hear this, yet he would not have been so pleased if he had realized the sheer impossibility of controlling the activities of hundreds of antiaircraft gunners spread across miles of front. It would only take one with an itchy trigger finger to set the whole lot off. In fact, because of either secrecy or sheer administrative incompetence, neither the naval gunners nor those manning guns on the merchant ships in the invasion fleet had been informed that there was going to be an airborne invasion, and so nobody was expecting to see friendly planes overhead. Even Admiral Hewitt, commanding the naval forces taking Patton's Seventh Army to Sicily, only found out about the airborne attack on the day it was due to take place. In such an administrative muddle, it is hardly surprising that the gunners were the last to learn.

On the evening of July 9, 1943, the British First Airlanding Brigade took off from six airfields around Kairouan in Tunisia. They were traveling in 137 Waco and ten Horsa gliders, piloted by men from the First Battalion of the Glider Pilot Regiment. The gliders were being towed by C–47s of Colonel Ray Dunn's American Fifty-first Troop Carrier Wing. Dunn's men were, in the main, pilots from civil aviation, and few had seen action or flown through flak. This proved to be a serious problem and contributed significantly to the catastrophe that was about to ensue.

As the C–47s and the gliders approached Sicily, a number of factors threw the entire air convoy into chaos. In the first place, the high winds made glider flying difficult, while the heavy flak coming up from Italian gunners forced the American pilots to take evasive action. To make matters worse, the gunners on the Allied invasion convoys joined in with the enemy gunners, and soon the sky was filled with planes and gliders going in every direction. In the words of historian Carlo D'Este: "The factors of inexperience, wind and enemy flak were a fatal combination and enough to ensure that Operation Ladbroke would be a first class disaster."

As they approached the shores of Sicily, the C–47 pilots made a decision that condemned hundreds of British airborne troops to a watery grave. According to Charles Whiting in *Slaughter Over Sicily*, the decision was taken to increase altitude to eighteen hundred feet, apparently to allow the gliders two free miles of flight once they were released from their tows. It also allowed the C–47s to release their gliders earlier than had been planned so that they did not need to fly through so much flak. Whatever the reasoning, the result was tragic. As Charles Whiting wrote:

> Some pilots decided that they had had enough. They did not even try to unload their tows. Instead they turned about and headed for North Africa again. Others released their gliders miles away from the landing zone and fled for safety . . . But the main fault, which led to the disaster to come, was that the gliders were released not two miles out to sea, but at twice or even three times that distance.[41]

Gliders now began to dive into the sea, miles from the shoreline. The glider carrying General Hopkinson, commander of the First Airborne Division and planner of Operation Ladbroke, was one of those to fall into the sea. Hopkinson managed to get clear of the wreck and hang on to a piece of wreckage until he was picked up by the British destroyer, *HMS Keren*. Ironically, Admiral Lord Ashbourne, flying his flag in the *Keren*, knew Hopkinson. As he wrote: "I saw a body floating in the sea almost alongside and evidently alive. I told the captain of the *Keren* to pick him up. A few minutes later a dripping soldier arrived on the bridge. He turned out to be Major-General G. P. Hopkinson, commanding 1st Airborne Division. The last time I had seen him was in 1922 when I had rowed in the same boat with him at Cambridge. We wrung out his clothes, gave him a plate of eggs and bacon and then sent him off to catch up with the rest of his soldiers."

Hopkinson was not going to find it an easy task to "catch up with the rest of his soldiers." The First Airlanding Brigade had been destroyed and scattered to the four winds. And Hopkinson thought he knew whom to blame—Colonel Dunn's tow pilots. Of the 147 gliders that had left Tunisia that evening, 69 had crashed into the sea, drowning 326 of the British "Red Devils." Of the others, just two had been shot down—by friend or foe—several had been towed back to Tunisia, and fifty-nine had landed somewhere in Sicily, though they were spread out across an area of twenty-five square miles. Resentment against the American tow pilots was so severe that when they got back to Tunisia, British troops there had to be confined to camp to avoid a lynching. In fact, two gliders even landed on different islands, one in Sardinia and another in Malta. The tragedies of war often have a comic side for those with a black sense of humor. The fate of one glider was truly farcical. Expecting to land on a rocky coastal area in Sicily, the men found themselves instead on a broad, sandy beach. Far from encountering Italian troops, or even German ones, the Red Devils were amazed to find that their whole invasion force had apparently been beaten to Sicily by a British mobile bath unit that passed them on the road. Then the truth dawned on them. They were not in Sicily at all; they had landed back in North Africa! Another glider crashed into the sea not far from a large, shadowy ship. The Red Devils aboard swam over to the ship and climbed up the anchor chain and onto the deserted deck. Suddenly, a sailor emerged on deck carrying a slop bucket. As he saw the commandos, he called for help, and soon the twelve Red Devils were being attacked by British sailors. Only with some difficulty could the twelve explain that they were not enemy saboteurs but glider troops, somewhat off target. Perhaps the most ridiculous of all the landings was by the glider that reached Malta. Its crew immediately organized themselves for action and got ready to move off toward their rendezvous point. Suddenly, a jeep drew up with two occupants who wanted to know what the Red Devils thought they were doing there. The young lieutenant leading the commandos replied that he was heading for his assigned LZ (landing zone) at Syracuse. A voice replied from the jeep: "We are sorry to inform you that you are not in Sicily, but on the main airstrip at Malta, and what's more, you are blocking one of the runways and the fighters cannot take off. So please take the jeep and pull not only the trailer but also this bloody glider 200 yards in that direction."

There was no disguising the fact that hundreds of British lives had been needlessly thrown away by the incompetence or cowardice of their tow pilots. The disaster had not been accidental: Mass hysteria on the part of inexperienced pilots had resulted in the ditching of gliders into the sea at night, miles from the shore and far from any chance of rescue. It had been a disgraceful and discreditable example of amicide.

While the massacre of the British glider troops was taking place, the second stage of the airborne attack was under way. Colonel James Gavin and the 505th Regimental Combat Team was already flying toward Sicily, following a path between the ships carrying Patton's American Seventh Army and Montgomery's British Eighth Army. Gavin was well-aware that if one of the naval gunners lost his nerve and opened fire on the hundreds of C–47s, a general massacre would ensue. But if Gavin's luck held while he passed over the ships, it began to run out when the air armada reached Sicily. As had happened to the British gliders, some of the American C–47 pilots panicked in the face of the heavy flak. The American paratroopers were scattered over a thousand square miles, many dropped at too low an altitude so that their chutes failed to open and they were killed outright or broke bones. Some pilots ordered their men out over the sea, while others simply turned around and flew straight back to Africa. The American paratroopers who fell into the British sector found in the early hours of July 10 that the British troops, not expecting to see Americans there, opened fire on them. Nobody had given the Americans the British passwords, and accidental casualties were common.

General Ridgway, from his floating headquarters, could only wince as news of the Gavin fiasco was added to the sorry tale of the gliders. None of the American troops had landed where they were supposed to, many being up to sixty miles from their LZs. Some had not even landed in Sicily at all, but were in Sardinia, Malta, or even—in a few tragic cases—in the mountains of Southern Italy, where their bodies lay undiscovered for many years.

If a more dismal fate was possible than that which had overtaken the British First Airlanding Brigade or Gavin's 505th Regiment, then it was about to happen to Colonel Reuben Tucker's 504th Regiment. While American and British infantry were coming ashore on the beaches of Sicily, General Patton was worried that few of Gavin's paratroopers were in position to hold up enemy counterattacks. He therefore ordered Ridgway to bring in the second part of the Eighty-second Airborne's contribution to Operation Husky—the 504th. Ridgway radioed to "Bull" Keenens in Africa to prepare the 504th for a drop that night. But the danger of friendly fire—so fearful a prospect to Matthew Ridgway—had increased during the day. The invasion fleet had been in constant action against German and Italian planes, and the flak gunners on the beaches and the ships had endured a harrowing day of almost relentless battle. German planes were continuing to stage hit-and-run attacks, while Stuka dive-bombers, sirens wailing, continually fell from the skies, spilling bombs onto the crowded decks of the transports. It was a day of high tension and bitterness. The British hospital ship *Talamba* was sunk by German bombs, and as night fell, the whole of the landing area was illuminated by the flames that had engulfed the American ammunition ship *Robert Rowan* as it was bombed. And into this scene of chaos soon would fly the slow C–47 transports carrying their precious cargo of American paratroopers. Their divisional commander, with a prescience of disaster, continued to press for guarantees of no naval fire. Earlier in the day, he had been told that the navy could make no such promises, whereupon Ridgway had warned Patton at Seventh Army Headquarters that unless he got his guarantee, he would "officially protest this follow-up drop." This appeared to do some good; Seventh Army gave Ridgway at least the impression that he had gotten his guarantee. In fact, he had not, because nobody could guarantee any such thing. Tucker's flight path would take him over thirty-five miles of invasion beaches, packed with Allied troops and defended by hundreds of antiaircraft guns, not to mention the hundreds more on the ships just offshore. With every gunner tired, nervous, blazing with anger at the loss of the hospital ship, how could anyone guarantee that they would not fire at the first planes they saw, planes that, for all they knew, were carrying bombs with their names on them. When Ridgway checked around the AA battery commanders, he found, to his horror, that more than one knew nothing at all about the order not to fire. An artillery liaison officer then reassured him that all would be taken care of at a briefing that afternoon. What more could Ridgway have done? He had to trust someone.

On July 11, Colonel Tucker in Tunisia had no reason to think that the drop would pose any special problems. With better weather than the previous night, and much lower wind, and with Gavin's men already—presumably—in position, his role was simply a backup. The flight to Sicily was straightforward, and aboard the lead C–47s, the paratroopers were preparing themselves for the drop. Behind them, the bulk of the regiment were being brought in at seven hundred feet over the massed ships, landing craft, and land troops of the American Seventh Army. Suddenly, the sound of a single machine gun was heard, firing up at one of the passing transports. It was like a signal for an ambush. It was followed by a few more desultory rounds from other guns, and then there broke out a roar as hundreds of guns began firing into the sky. The C–47s desperately fired recognition flares, but they were answered with a deluge of bullets, shells, and rockets. When the gunners were later questioned, most simply explained that they thought the low-flying transports were German bombers and that they had taken so much punishment during the day that they were looking to get some of their own back. Some of the more inexperienced men admitted that they could not recognize a C–47 from a German bomber and others claimed they thought the planes were dropping German paratroopers onto the beaches. In fact, it was one of history's clearest examples of mass hysteria. So tight was the tension felt by all the combatant soldiers on the Gela beaches that night that a single incident could make a hundred—a thousand—men act as one, without rational thought. With the C–47s flying at just seven hundred feet, it was almost impossible for the gunners to miss. Aboard the transports, paratroopers were killed in their seats, shot by machine-gun bullets or ripped by shrapnel. Below, watching the slaughter, Ridgway and Patton stood staring in horror: Ridgway in tears for the massacre of his fine division, Patton simply repeating: "Oh, my God! Oh, my

God!" But there was no divine intervention that evening. Tucker's men were beyond salvation. Planes were crashing into the sea, or onto the beaches. For days afterward, the bodies of paratroopers were washed up at Gela.

During the initial drop, 33 out of the original 144 C–47s were shot down in minutes, while another 60 were so badly hit that they would never fly again. A total of 318 paratroopers were killed or wounded, over one in five of the men involved in the operation. And all were victims of friendly fire. Captain Adam Komosa later described his experiences:

> It was the most uncomfortable feeling knowing that our own troops were throwing everything they had at us. Planes dropped out of formation and crashed into the sea. Others, like clumsy whales, wheeled and attempted to get beyond the flak which rose in fountains of fire, lighting the stricken faces of men as they stared through the windows.[42]

Colonel Tucker was one of the fortunate ones who landed at the correct dropping point. As he touched down, he found, to his horror, that not far away a group of turret gunners in Sherman tanks were firing wildly into the sky at the passing planes. Tucker's meeting with these men can better be imagined than described. When the plane that transported the colonel got back to Tunisia, it was found to have been holed over a thousand times by bullets and shrapnel, all of which was fired by American guns. But now the perpetrators of this shocking amicide let their madness take them over the line that separates war from atrocity. It seems that the naval gunners, exhilarated by this massacre of what they believed to be German planes, now began a "turkey shoot," killing the parachutists as they helplessly floated down, many of them leaping only to escape from the burning C–47s above. The night sky was illuminated by the burning planes and parachutes, and by the tracer fire arcing upward. One witness of that night, Herbert Blair, spoke poignantly of what he saw:

> Hit after hit we scored until ship [C–47] after ship bursts into flame or falls spiralling into the sea. But something is wrong. From the wounded ships parachutes come fluttering down,

> some in flames, others to billow out in a slow descent. Then some trigger-happy gunner aboard another ship decides to pick off the supposedly helpless Jerries. Soon every gunner is firing at the troopers who dangle beneath the umbrellas of their chutes . . .
>
> "Cease firing! Cease firing! Stand buy to pick up survivors! Stand by to pick up survivors!"
>
> Only then does the dreadful realization descend like a sledgehammer upon us. We have wantonly, though inadvertently, slaughtered our own gallant buddies. I feel sick in body and mind.[43]

An American destroyer, the *Beatty*, fired on a plane that had ditched in the sea and continued firing for quite some while before it was recognized as American and boats were launched to rescue the survivors. With the grimmest of all grim humor, one of the C–47 pilots remarked later: "Evidently the only safe place for us over Sicily tonight is over enemy territory."

By the next morning, the full extent of the catastrophe was clear. Of a total of 5,307 paratroopers who had flown in the previous day, fewer than 2,000 were still fit for combat, and more than 1,200 of them had died from acts of amicide. General Eisenhower was shocked by the outcome of the whole airborne operation. Essentially, the elite troops of the American Eighty-second Airborne Division and the First British Airborne Division had been frittered away as the result of what came to be known as the "Sicily Disaster." Ike wrote to Patton: "If the cited report is true, the incident could only have been occasioned by inexcusable carelessness and negligence on the part of someone. You will institute within our command an immediate and exhaustive investigation into the allegation with a view to find responsibility. Report of pertinent facts is desired and if the persons found responsible are serving in your command, I want a statement of the disciplinary action taken by you. *This will be expedited.*"

How could Eisenhower possibly account for the loss of so many Allied lives when the enemy had not even been involved? There had been a scandal, and he wanted a scapegoat. But where could one begin? Certainly not with the green gunners who panicked after a day's aerial battering by Stukas.

Even as the search for scapegoats was initiated, the last phase of the ill-fated airborne assault was under way: This time the elite British First Parachute Brigade, commanded by Brigadier Lathbury, was on its way to Sicily. In terms of casualties, Lathbury's men did no better than their predecessors, for chaos was still the order of the day. The commander of the First Battalion, Colonel Pearson, had to force his American pilot to fly on through the flak by holding a revolver in his back and threating to shoot him after the man had tried to turn back. But when the time came to jump, many of his men were killed by enemy fire as soon as they jumped from the transport. The Third Battalion, under Colonel Yeldham, was just as unlucky, losing many men to enemy fire and, as he recorded in his diary, to "the fire of our flak ships." Of sixteen gliders that accompanied Lathbury's flight, four were shot down by friendly fire and most of the others succumbed to enemy guns. Of Lathbury's paratroops, most went the same way that Gavin's "All-Americans" had gone, scattered to the winds, some ending up on the Italian mainland but far too many others not reaching land at all and perishing in the dark waters of the Mediterranean. One group even landed on Mount Etna, though as many as thirty percent were taken straight back to Tunisia. To fight the full strength of two German regiments, with Tiger tanks, Lathbury found himself at the dropping zone with just 295 men out of his total force of 1,856.

If Eisenhower was looking for someone to blame, he should have started with whoever planned the transport for the airborne troops. The poorly trained American pilots who flew the C–47s, many of whom were airline or freight pilots in civilian life and had no experience with such tough and dangerous work, were guilty of at best funk and at worst a scandalous disregard for the safety of their passengers. Both in the British and American sectors the pilots were responsible for the death of hundreds of paratroops, most simply ditched into the sea at night. But so disgraceful were the facts when they reached him that Eisenhower decided the whole fiasco must be hushed up. There were too many reputations at stake. As a result, it was interesting to note the skill with which the buck was passed when the investigations began. The officer commanding the troop-carrying operation placed the blame fairly and squarely on the shoulders of the army and navy antiaircraft gunners, who frankly panicked at the sight of so many low-flying planes. This view was supported—unsurprisingly—by air commanders involved, like General Spaatz, head of the USAAF in the Mediterranean. The shooting down of Tucker's planes, according to Spaatz, was the result of a breakdown in coordination among all three services. To spread the guilt even further and more thinly, Spaatz blamed command decisions at headquarters in Malta, Tunisia, and even Egypt. No attempt was made to find out who had fired first on the night of Tucker's martyrdom, whether army or navy gunner.

The real cause of the problem was the fact that the airborne troops were there at all. And at this point, Eisenhower's deputy, Air Chief Marshal Lord Tedder, entered the fray. Tedder's well-known antipathy toward Montgomery was clearly at the root of his criticism of the whole operation. It had been too dangerous, according to Tedder, for the airborne troops to be carried across an area of open battle. Tucker's men had to fly over thirty-five miles of beachhead, where the troops had only just suffered air attack by the enemy. It was far too risky to expect men in such a state of tension not to respond to such a potential danger as Tucker's C–47s must have seemed to offer. It was risking amicide, and nobody should have been surprised when the paratroops suffered disaster. In a way, Tedder was right. After all, General Matthew Ridgway had been worried about friendly fire from the beginning and had pressured his fellow commanders to guarantee that the gunners would not fire at his men. This they proved unwilling to do for a reason that Tedder was at lengths to explain. The British air chief marshal wrote: "Even if it is physically possible for all the troops and ships to be duly warned, which is doubtful, any fire opened up either by mistake or against enemy aircraft would almost certainly be supported by all troops within range. AA firing at night is infectious and control almost impossible."

Admiral Cunningham, Britain's finest fighting sailor of the Second World War, was less than helpful. As far as he was concerned, it would have been far better to keep friendly planes away from his ships. He had made it quite clear from the start that he would not guarantee that his guns would not fire at British and American planes for the simple reason that he could not risk the safety of a ship on a rapid identification of a distant aircraft. He applied the rule that his gunners would fire at any aircraft that flew too near. If the planners had selected better routes and if the pilots had navigated better, the disaster need not have happened. With the navy blaming the army planners, it was incumbent on the latter to reply and they duly did through Lieutenant-General Sir Frederick Browning, who returned the blame to the pilots: "The navigation by the troop carrier aircrew was bad . . . It is essential both from the operational and morale point of view that energetic steps are taken to improve greatly on the aircrew's performance up to date." But Eisenhower was not having any of this. He agreed with the view that the operation was just too dangerous, and he prohibited any further airborne operations by the Eighty-second Division. But Montgomery felt that there was still a future for British airborne operations, provided that the transport system and the pilot performance was improved. The prickly and chauvinistic British general was notably blunt about the American pilots: "The big lesson is that we must not be dependent on American Transport aircraft, with pilots that are inexperienced in operational flying. Our airborne troops are too good and too scarce to be wasted."

But the final word was left with Ridgway. It was elegantly expressed, but it was a lie. "The lessons now learned could have been driven home in no other way, and these lessons provide a sound basis for the belief that recurrence can be avoided. The losses are part of the inevitable price of war in human life."

It was also during the conquest of Sicily that American aircraft first earned the uncomplimentary epithet from the GIs of the "American Luftwaffe." In Sicily, the North American A–36 Invader, admittedly a first-rate dive-bomber, had a depressing record of hitting friendly troops by mistake. During the progress of the Second Armored Division's pursuit of the Fifteenth Panzer Grenadier Division, American planes made repeated attacks on their own troops, killing and wounding seventy-five men. Even when the ground forces displayed luminescent identification, this did not seem to stop the A–36 once it had started its dive. General Omar Bradley narrowly missed falling victim to an A–36, which dive-bombed and strafed him while he was visiting General Allen's HQ. Bradley observed that he was getting used to it; it was his third strafing that day by American planes. The A–36s were also responsible for the loss of Monte Cipolla to the Germans. A group of American GIs were desperately holding a position on the mountain until seven A–36s bombed them, killing or wounding nineteen men, destroying their last four howitzers, and driving them headlong down the slopes. In the same area, near the town of Troina, A–36s also dive-bombed British Thirtieth Corps Headquarters, mistaking it for a German-held position. The attack was actually witnessed by General Omar Bradley himself:

> For three days [General] Allen's attack on Troina was thrown back by savage resistance . . . Troina itself was to be bombed until it surrendered or was smashed into dust.
>
> On the late afternoon of 4 August I waited at a bend in the road, high up in Cerami, to witness this air attack, the heaviest to date in our Sicilian campaign. Across the bowl-like depression, now half obscured in dust, the fire from eighteen battalions of artillery hammered the enemy's AA positions.
>
> Thirty-six fighters circled high overhead, each loaded with 500-pound bombs. The artillery slackened and the bombers peeled off in a near-vertical dive. Soon the crown of Troina was wreathed in dust. By the time a second flight of thirty-six planes had bombed that stricken city, Troina lay half obscured under a column of grey dust that partially hid the cone of Mount Etna. Once more the infantry started forward, but once more the enemy held and slashed back in counter-attack.
>
> The following day we renewed the offensive. This time Major-General Edwin J. House, Patton's tactical air commander, accompanied me to Cerami to view the air bombing. H hour passed with no sign of air. As we were about to leave in dismay, a drone sounded far off to the south. There, high in the sky, three A–36s were high-tailing for home.
>
> "Holy smokes," I turned to House, "now just where in hell do you suppose they've dropped their bombs?"
>
> "I'll be damned if I know," he said.
>
> "Maybe we'd better get back to your headquarters and see what went wrong."
>
> On our arrival the phone was ringing. It was Oliver Leese from British XXX Corps.

"What have we done that you chaps would want to bomb us?" he asked.

"Where did they hit?" I groaned.

"Squarely on top of my headquarters," he said, "they've really plastered the town."[44]

Eventually, when A–36s shot up an American tank column in spite of yellow recognition signals, the GIs lost patience and shot one of them out of the sky. The pilot parachuted down and was furious when he found out that he had been shot down by friendly troops. "Why you silly sonuvabitch," said the tank commander, "didn't you see our yellow recognition signal?" "Oh," said the pilot, "is that what it was?"

Even after the Allied troops had crossed the Straits of Messina into mainland Italy, American airmen continued to exact an alarming toll on friendly troops. Heavy bomber strikes caused widespread havoc, inflicting heavy friendly casualties at Venafro, during the bombing of Monte Cassino. Venafro was fifteen miles from the target area, and the American planes managed to destroy the British Eighth Army commander's caravan (the long-suffering General Leese was fortunately not in it) and a Moroccan military hospital, causing 150 casualties among the civilians, as well as among gunners from the Fourth Indian Division. During the advance on Rome, American Mustang fighter-bombers strafed columns of American troops by mistake, inflicting hundreds of casualties. American General Mark Clark was furious about the Venafro fiasco, during which fifty-seven Allied soldiers and friendly civilians were killed, putting it down to "poor training and inadequate briefing of personnel." But a New Zealand officer was more perceptive: "Heavy bombers from 14,000 feet are not accurate enough for this class of close support. Medium and light bombers are excellent." If only Omar Bradley could have come to the same conclusion, the Cobra tragedy might have been avoided.

Operation Cobra and Afterward

Far from promising a prompt end to the war against Germany, the D-Day landings in Normandy in June 1944 seemed to offer a prospect not far different from the dreadful static warfare of 1914–18. British and American planners, possibly blinded by their enormous material advantage over the Germans, notably in airpower, had blundered in underestimating the problems posed by the difficult Normandy terrain, particularly the bocage, the thick hedgerows that slowed down even the Allied tanks. At the end of the war General Bedell Smith admitted:

> All commanding officers were theoretically aware of the hedgerow terrain, but none had seen it. You cannot imagine it when you have not seen it. I had seen air photos of it but I could not imagine what it was like. Field Marshal Brooke, who had fought there, was very pessimistic about our chances.[45]

This bocage was skillfully used by the German defenders to slow up the Allied advance, and the unusually stormy weather in early July 1944 hampered Allied airpower from supporting their troops. In fact, the whole huge enterprise that had begun on June 6 was being held up by the slow progress out of the bridgeheads. Huge numbers of troops, with all their heavy equipment as well as their air support, were waiting in England and in the United States to ship to France but there was literally no space ashore to take them all.

There was a clear need for a breakout from the Normandy bridgeheads, and among ground commanders, the belief was that heavy bombers acting as ground-support aircraft could carpet-bomb sections of the German front lines to create passages for the Allied troops. Such use of heavy bombers would require close liaison between ground and air commanders, yet the pressures of the war situation were such that this close relationship was never possible. The main point at issue was over the role of the heavy bomber, which had never been designed for ground support. In the eyes of the air commanders, heavy bombers were a vital part of the strategic bombing campaign against Germany, and in this role alone, they would contribute to Allied victory. The transfer of such bombers to ground support was extremely ill-advised if not actually counterproductive. Certainly their capacity to carpet-bomb was not in question, but their capacity to bomb precisely at a time when friendly troops were within range was a very different matter. As supreme Allied commander, General Eisenhower was under pressure to use everything available to get his ground troops moving again, and if heavy bombers could assist, then he was prepared to use them. But there was a fundamental disagreement between the Allied air commanders. The deputy supreme commander, British Air Chief Marshal Arthur Tedder, backed by United States Strategic Air Force commander Lieutenant General Carl Spaatz, was at odds with SHAEF air commander Air Chief Marshal Trafford Leigh-Mallory. Tedder and Spaatz had no intention of handing over their heavy bombers to Eisenhower and Leigh-Mallory for their ground-support operations. They both doubted Leigh-Mallory's strategic grasp of the war, and Spaatz in particular was no fan of the British airman's judgment. Nevertheless, Leigh-Mallory retained Eisenhower's confidence, and he continued with his plans for carpet bombing to precede an Allied breakout from the Normandy bridgeheads, threatening to resign if Tedder or Spaatz tried to stop the operation. The situation favored Leigh-Mallory in that ground operations had virtually come to a halt. Something was needed to regain the impetus of the Anglo-American invasion.

The plan for carpet bombing was targeted on General Omar Bradley's First Army front at St. Lô and was to be known as Operation Cobra. The aim would be to carpet-bomb an area in front of Seventh Corps to enable it to break out. By July 19, Bradley's troops had taken St. Lô and were ready to unleash Operation Cobra, but first he needed to win the close and precise cooperation of the air chiefs in England. What Bradley wanted was something that had never been tried before in war: the sort of saturation of the enemy front in sixty minutes that had taken thousands of British guns days and weeks to achieve on the Somme in 1916 and at Passchendaele the following year. The bombers would concentrate on an area five miles wide and one mile deep, using light bombs to achieve minimum cratering and maximum antipersonnel effect. He hoped that the impact on the German defenders would be so tremendous that they would be incapable of offering much resistance as Seventh Corps drove through them. Unfortunately, professional jealousy among the Allied leaders was rife. The airmen regarded Bradley as a dilettante in air matters and felt that his plan was unrealistic. Heavy strategic bombers simply did not have the tactical capacity he was demanding. Bombing so close to friendly forces was just asking for trouble. They wanted a safety zone of at least three thousand yards, and even this was scarcely enough to guarantee that there would be no mishaps like friendly casualties. But a three-thousand-yard safety zone was ridiculous in Bradley's eyes. What was the use of pounding the enemy until he was dazed and then positioning your assault troops so far back that by the time they had crossed the safety zone, the Germans had recovered and were ready to repel the assault? To Bradley, a distance of eight hundred yards was the maximum he would ask his men to fall back. To the airmen, this was tantamount to suicide. And so the dispute dragged on; airmen and footsloggers might as well have been speaking a different language. One group knew the realities of air warfare, the limitations in bombing technique of heavy bombers designed to blast areas rather than precise targets, and the near certainty of friendly casualties if the safety margins were cut too fine. The other group were versed in the problems of land warfare, in which to hand over three thousand yards of hard-won ground to the enemy was unacceptable. Eventually, a compromise solution was reached that satisfied neither side. Bradley would order his men to fall back fifteen hundred yards, and the airmen would try to operate within the constraints of an infantryman's scenario. It was a formula for catastrophe.

Omar Bradley had given a lot of thought to how the heavy bombers should approach the area to be bombed. Aware of the dangers of friendly casualties, he wanted them to come in parallel to the road from St. Lô to Periers, so that they did not overfly his own ground troops. To Bradley—on the ground—the road was the most important and the most obvious feature of the area. However, from fifteen thousand feet or so, the airmen did not consider the road as prominent a feature as Bradley felt it to be. To make matters worse, experienced airmen saw obvious flaws in Bradley's plan to fly in parallel to the road. With a time limit of just sixty minutes, it was stretching the bounds of possibility to expect fifteen hundred heavy bombers to fly in over an area just one mile wide. The air chiefs insisted that only a north-south approach, using the Normandy coastline for reference and the St. Lô–Periers highway as a sighter for the bombardiers, was feasible for an operation on such a large scale. And so the matter stood, with Bradley insisting on a parallel approach to the road, with the bombers overflying German lines, and the air chiefs insisting on a north-south approach, with the bombers overflying American troops. The problem was that at the meeting at Stanmore, in England, Bradley and air chiefs like Leigh-Mallory failed to appreciate the substantial differences that existed between them. Bradley left assuming that the airmen agreed with him, and after the fiasco that followed on July 24–25, he accused them of deliberately misleading him. This was unfair, for he was asking them to do something that they considered impossible.

Preparations for Operation Cobra now speeded up, and the attack was earmarked for July 21. Seventh Corps troops were instructed to "vigorously push the attack across the highway to insure annihilation of any remaining enemy." The emphasis on "any remaining enemy" was significant and shows the confidence the ground commanders had in the efficacy of the proposed bombing. It was supposed that the Germans would be so shattered by the bombing that few would survive and those who did would offer only feeble resistance to a determined American push. However, ground commanders were less happy to receive orders to withdraw nearly a mile to offer a safety zone. This mile had been fought over and won with the lives of their soldiers.

Had Bradley heard the instructions that were being given to air force bombardiers, he would have been gravely concerned. They were instructed not to bomb short "because the penetration route is directly over friendly troops." This would have alerted him to the fact that his insistence on a parallel approach by the bombers to avoid friendly casualties had been rejected in favor of a north-south approach, at right angles to the St. Lô to Periers highway. The misunderstanding between Bradley and the air chiefs was going to cost a high price in American lives.

Operation Cobra involved a vast commitment in airpower by two tactical air commands, Ninth TAC and Nineteenth

TAC. At the start, fighter bombers would attack a strip along the St. Lô to Periers highway, to be followed by an attack from 1,586 heavy bombers, coming in over the English Channel at a height of some fifteen thousand feet. These Eighth Air Force "heavies" were to attack in three waves, each lasting fifteen minutes, with a five minute interval between waves. The target area would be hit by fifty thousand general purpose and fragmentation bombs, most of relatively small size—one-hundred-pounders. Formations of twelve to fourteen bombers would follow the example of a lead aircraft and drop their bombs in accordance with the lead bombardier. As if that was not enough, once the heavies had delivered their payloads, German rear areas were to be attacked by medium bombers from the U.S. Ninth Air Force. It would be a prodigious display of Allied air mastery, and according to the standards of the time, it should have been adequate to achieve the saturation of German defenses that Bradley was seeking. But with operations on such a scale, some things were bound to go wrong. The first problem was the weather. On July 23, heavy cloud made flying impossible, and an irate Bradley commented, "Dammit, I'm going to have to court-martial the chaplain if we have much more weather like this." But the following day, Leigh-Mallory overruled another Eighth Air Force request for a postponement and gave the order to begin.

In the last few days before Cobra began, it was obvious that there was still general confusion among the air chiefs about the approach to the target area to be followed by the Eighth Air Force heavy bombers. Leigh-Mallory and his deputy, Hoyt Vandenberg, continued to question the planners involved in bringing in the heavies. They were unhappy to learn that the Eighth Air Force men considered Bradley's plan for a parallel approach impracticable; there just was not enough time to bring in more than fifteen hundred bombers along that path. Vandenburg was told by Major General Fred Anderson that "he was worried about the repercussions that might arise and that he wanted it clarified that the time factor which was sent by AEAF was the controlling one for their direction of attack." Vandenburg told Leigh-Mallory that Bradley's plan was a non-starter and that the general might prefer to extend the time beyond one hour to allow the parallel approach that he preferred. But the British air chief replied that he had already discussed this with Bradley and he was unwilling to extend the bombing time. According to Leigh-Mallory, Bradley "had decided to accept the additional risk of perpendicular to the road bombing." This statement was later vehemently denied by Bradley. But in view of the catastrophe that followed the saturation bombing, nobody was in a hurry to accept responsibility, and it is doubtful that we will ever know where the truth lies.

Heavy cloud on July 24 postponed the operation by the fighter bombers, but by the time Leigh-Mallory decided to call the whole show off, it was already too late to prevent many of Eighth Air Force's heavy bombers from taking off. The Second Bomber Division, however, aborted its own attack as heavy cloud cover over the target area made identification impossible. A minor incident when a single bombardier accidentally unloaded his bombs on an Allied air field was not too serious a problem for such a large operation. The Third Bomber Division also failed to locate the target, though perhaps forty planes did drop their bombs in the vicinity. The real problems began when the First Bomber Division arrived over France and found that conditions were improving. Although some of these aircraft received the recall signals, most did not, and 317 heavy bombers dropped a total of 10,124 high-explosive bombs and 1,822 fragmentation bombs. In the confusion, some of these bombs fell short. Individual problems—one lead bombardier had a faulty bomb release mechanism—were magnified many times over. This unfortunate airman, by dropping short, was followed by a further twelve or fourteen planes that took their lead from him. On the ground, men from the U.S. Thirtieth Division were already diving for cover in ditches. In seconds, 25 soldiers were killed and a further 131 wounded. The Cobra fiasco had begun.

General Omar Bradley was shocked when the news reached him. It was exactly what he had feared and what he had tried so hard to avoid. How had it been possible? When he realized that the bombers had flown in on a perpendicular line rather than the parallel one on which he insisted, he was furious. The airmen had lied to him. He asked Leigh-Mallory for an explanation. Leigh-Mallory promised to get one—from Eighth

Air Force. But, according to Vandenberg, the matter had already been discussed some days before. Was Leigh-Mallory really ignorant of Eighth Air Force's intentions to fly in perpendicular rather than parallel? Again it seemed that someone was lying.

On the ground, news of the postponement gave the Seventh Corps commander, General Lawton Collins, no real alternative but to reoccupy the safety zone that he had abandoned only hours before. There were a lot of disgruntled GIs after the zone had been recaptured at the expense of some casualties. At least the Germans were left with the impression that the day's fiasco meant that a major American attack had been repulsed.

Meanwhile, Leigh-Mallory had decided to try again the next day—July 25—when the meteorologists promised better weather. But first he had to persuade Bradley to accept another perpendicular approach by the heavy bombers. As he pointed out, it was now far too late to change Eighth Air Force's mind about the time needed for a parallel run, and so—uneasily—Bradley accepted. The operation was now timetabled for 0900 on July 25.

On the morning of July 25, the weather had improved, but there was still cloud cover at fourteen-thousand feet, making it necessary for the heavy bombers to fly at a lower altitude than planned and requiring the bombardiers to recalculate their bombing data. In itself, this should not have posed many problems, but in an operation on such a vast scale, small errors were likely to have great effects.

The first problem was that a wind coming from the south blew a huge column of smoke from the target zone—which had just been hit by Ninth Air Force's fighter-bombers—northward, directly into the path of Eighth Air Force's armada of 1,495 heavy bombers, B–17s, and B–24s. Even without the problem of cloud cover, it was extremely difficult for navigators and bombardiers to pick out landmarks, for the whole area was covered by immense clouds of smoke, dust, and red marker flares. Nevertheless, the American bombers succeeded in dropping over four thousand tons of bombs into and around the target area. Errors were inevitable, and they were on a scale that matched the entire operation. According to the official report by Walter E. Todd, human error contributed to extensive friendly casualties. Two lead bombardiers released their bombs without first achieving satisfactory identification and were then followed by their entire units, while a command pilot caused short bombing when he assumed his wing was supposed to bomb as a single unit. The results, on the ground, were catastrophic. The Thirtieth Infantry Division, which had suffered casualties the previous day, was again in the forefront of the blue-on-blue. A total of 61 men were killed, 374 were wounded, and 64 others were listed as missing, and 164 men suffered total nervous collapses as a result of shell shock or "combat fatigue." Other parts of the Seventh Corps also suffered losses, and included in the final list of 111 fatalities was General Leslie McNair, the most senior American officer lost during the entire war. McNair had been in a forward position—despite warnings that he should stay at the rear—observing the effects of the saturation bombing. He died when his bunker suffered a direct hit. Bradley was horrified when he received the news. "Oh Christ," he said, "not another short drop." War correspondent Ernie Pyle was with the troops on the ground and recorded the shattering effects of the bombing:

> As we watched there crept into our consciousness a realization that windrows of exploding bombs were easing back towards us, flight by flight, instead of gradually forward, as the plan called for. Then we were horrified by the suspicion that these machines, high in the sky, and completely detached from us, were aiming their bombs at the smokeline on the ground, and a gentle breeze was drifting the smokeline back over us! An indescribable kind of panic comes over you at such times. We stood tensed in muscle and frozen in intellect, watching each flight approach and pass over us, feeling trapped and completely helpless.[46]

Lieutenant Colonel George Tuttle of the Thirtieth Division described his experiences: "The ground was shaken and rocked as if by a great earthquake. The concussion, even underground, felt as if someone was beating you with a club." Young Lieutenant Sidney Eichen of the 120th Infantry, like many other American soldiers that day, felt proud as he saw the seemingly endless lines of American planes flying toward him. Their power seemed limitless. Suddenly, he felt the awful realization that, just as they had yesterday, these bombers were going to drop short over American lines. As he wrote afterward: "My outfit was decimated, our anti-tank guns blown apart. I saw one of our truck drivers, Jesse Ivy, lying split down the middle. Captain Bell was buried in a crater with only his head visible. He suffocated before we could reach him." The bitterness of the GIs toward the airmen who killed them knew no bounds, and many of them fired their rifles impotently into the skies at the planes that had earlier filled them with such pride.

Nevertheless, in spite of the appalling casualties to friendly forces, the bombing went on and was immediately followed up by American infantry assaults on German fronts. Disappointingly, the first waves of American troops found the Germans, apparently unruffled by their ordeal, waiting for them, and resistance was firm. It transpired that German losses were only marginally heavier than those of the Americans—700 to 601—and was a good advertisement for the effectiveness of German tunneling. Unlike the Americans, few of whom had dug foxholes, the Germans had been thoroughly prepared for anything less than the Day of Judgment. On the other hand, American troops can be excused for not expecting to need such protection against their own aircraft.

German resistance on the afternoon of July 25 was unexpectedly strong. A German commander, Fritz Bayerlein, explained that although German forward positions had been destroyed, reserves had been held back beyond the target area and were quickly rushed to the front once the bombing had stopped. As a result, American progress was slow, and when the First Army failed to achieve the expected breakout, the search began for scapegoats. The army blamed General Doolittle's Eighth Air Force for "lacking enthusiasm for ground support." But the air chiefs responded by blaming Bradley. He had been warned that if he reduced the safety zone below the three thousand yards that had originally been demanded as the minimum safe distance for heavy bombing, friendly casualties were almost inevitable. As General Spaatz said: "We were attempting to place too heavy a concentration in too small an area."

But all was not as it appeared. German defenses were a facade. General Bayerlein described the experiences of the German troops on July 25:

> It was hell . . . The planes kept coming overhead like a conveyor belt, and the bomb carpets came down, now ahead, now on the right, now on the left . . . The fields were burning and smouldering. The bomb carpets unrolled in great rectangles . . . My front lines looked like a landscape on the moon, and at least seventy per cent of my personnel were out of action—dead, wounded, crazed or numbed. All my front line tanks were knocked out. Late in the afternoon, the American ground troops began filtering in. I had organized my last reserves to meet them—not over fifteen tanks, most of them from repair shops. The roads were practically impassable. Then next morning the bombing began all over again. We could do nothing but retreat.[47]

If the Americans pushed hard, the whole edifice would collapse, and they would get their breakthrough. General Collins had already decided to fling in his armor on the morning of July 26, and this would prove decisive. Cobra would be a success after all, but one achieved at the price of the worst aerial blue-on-blue in history.

Who was to blame for the dreadful Cobra fiasco? Omar Bradley was convinced that he knew the answer. It was the air chiefs of the Eighth Air Force, who had refused his reasonable plan for a parallel approach over German lines and instead opted for the perpendicular approach that took the planes over friendly troop concentrations. His words were bitter, even libellous. "It was duplicity," Bradley wrote in his autobiographical *A General's Story,* "a shocking breach of good faith." But this was untrue, and Bradley must have known that even as he wrote the angry words. He had been told repeatedly that sixty minutes was far too short a time for fifteen hundred heavy bombers to carry out the operation if they approached parallel to the St. Lô to Periers highway. Nor was it certain, even if they did approach in that way, that no bombs would fall on friendly troops. A safety margin of three thousand yards was the minimum the airmen thought advisable, and Bradley was being unrealistic by reducing it to eight hundred or even the fifteen hundred yards that was the final unwilling compromise. Bradley simply did not understand and was apparently unwilling to try to understand the problems of the strategic bomber asked to become a tactical bomber for a "one-off" mission. He was expecting the flyers to bail out his First Army at a low point in their fortunes.

If blame was to be apportioned, Bradley had to be prepared to shoulder his share. Aware of the dangers from blue-on-blue bombing, he still made no attempt to take safety measures to minimize the danger when the time came for the saturation bombing. Why were the American troops out in the open, cheering on the "heavies," rather than in foxholes and trenches like the Germans? The simple answer, of course, is that he was relying on being able to attack the Germans straightaway, while they were still groggy from the bombing. To have dispersed the Seventh Corps and sent them under ground would have used up vital minutes when the time for the infantry assault came. This was a senior command decision. Friendly casualties might need to be risked to reduce casualties from enemy fire. It was an equation to be balanced by the officer commanding, and it reflected little credit on Bradley that when he got his sums wrong, he tried to blame someone else for his own failings.

In terms of blame, Bradley did not stand alone. Air Chief Marshal Trafford Leigh-Mallory failed in his task to coordinate ground and air units during Operation Cobra. His own experience had been with single-seat fighters, and he really knew little and understood less of the problems of the heavy bomber. In his liaison capacity between the British and the Americans, Leigh-Mallory was something of a disaster, and even a British officer at SHAEF wrote him off as having a reputation for incompetence and "a pompous, arrogant attitude." His decision to let the "heavies" set off on July 24, even overruling in the process the opinions of the weather expert, before recalling them, too late in some cases, contributed to the tragic friendly casualties on that day. In addition, it also meant that the American troops who had abandoned the safety zone on July 24 had to fight to regain the territory when news came that the operation was postponed. A further, extraordinary decision that formed part of the Cobra bombing was the use of Ninth Air Force's fighter bombers as a preliminary to the work of the heavy bombers from Eighth Air Force. The obvious problem—and one that should have been obvious at the time—was the obscuring effect of all the smoke from the fighter bombers' ordnance. With a wind from the south, this smoke would be blown across the approach paths of the "heavies," with the result that target identification would be difficult, the St. Lô to Periers highway invisible, and careless bombing much increased. It was a disastrous formula, particularly for airmen flying over territory occupied entirely by friendly forces. In such a case, blue-on-blue was not so much a possibility as a certainty.

In August 1944, the Canadians launched an armored offensive against German troops at Caen. As in the case of Operation Cobra, this operation—code-named Totalize—was to be prefaced by an aerial bombardment of German positions. On August 7–8, one thousand bombers from RAF Bomber Command were due to saturate the forward German positions, but such was the buildup of smoke and dust that at least a third of the planes did not drop their bombs for fear of hitting friendly troops. The next day, the U.S. Eighth Air Force took over ground support and unfortunately inflicted three hundred casualties on the Canadians and Poles. This error was followed up by the RAF and RCAF on August 14, during its support for the assault on Falaise, code-named Operation Tractable. At first, all went well until seventy-seven Lancaster and Halifax bombers of the second wave flew over the Canadian and Polish lines, dropping their bombs on friendly troops, causing sixty-five deaths and wounding over four hundred men. By an incredible oversight, nobody had informed bomber command that ground troops were identifying their positions by yellow smoke: The fact was that yellow smoke was bomber command's target-identification color. The more desperately the ground troops burned their yellow flares, the more the bombers rained death on them. Worse casualties were only avoided when an Auster trainer took off and flew in front of the bombers, waggling its wings and trying to draw them away from the Canadian lines.

Aerial amicide continued to exact a heavy price during the late summer of 1944. So extensive had been the friendly casualties during the massive heavy bomber raids in Operations Cobra, Totalize, and Tractable that the use of heavy bombers was discontinued and ground support was left to the more flexible medium bombers and fighter-bombers. However, although this decision reduced the severity of the incidents, it did not necessarily reduce their number. The very scale of operations and the increasing fluidity of the battlefield that followed the Allied breakouts from Normandy contributed to a rash of minor friendly fire incidents. Human errors among pilots and navigators led to numerous strafings and bombings of Allied troops so that few units even bothered to record them all. Among the ground troops, it was merely a sign that they had a new enemy in the sky—the American Luftwaffe. As usual, the Thirtieth Division was in the thick of it. On July 29, near Troisgots, American fighter-bombers bombed and strafed several units from the division. A few days later, during the heavy fighting around Mortain, the division was frequently attacked by American P–47s and British Typhoons. On August 7, the 120th Infantry recorded ten separate incidents of friendly fire by Allied fighter bombers, while two U.S. tanks from the Third Armored Division were destroyed by friendly aircraft. Near Laval, American fighters shot up Third Army and Nineteenth Tactical Air Command headquarters. Tired of taking this aerial punishment, antiaircaft gunners around Laval promptly shot down one of the attacking planes.

The fact that Anglo-American troops were having to fight their way through France, Belgium, and Holland to liberate them from the Germans involved the obvious difficulty of trying to destroy a tenacious enemy without harming the friendly population of the areas through which Allied troops moved. And when one adds the extra difficulty of using close air support and tactical bombing to support the ground troops, the chances of friendly casualties were much increased. On October 2, 1944, American planes made "a gross error of navigation" and attacked the Belgian town of Genck, thirty-five miles from the intended target inside Germany. The result was that thirty-four Belgian civilians died and forty-five others were injured.

Just before Christmas 1944, the Belgian city of Malmedy suffered further attacks by American planes. The apparently ubiquitous Thirtieth Infantry Division was in combat with the First SS Panzer Division near La Gleize. Ordered to support the ground troops, six B–26s missed their designated target, which was Zulpich, the railhead of the German Seventh Army, and instead dropped their bombs on Malmedy, killing at least thirty-seven soldiers of the Thirtieth Division and many Belgian civilians, and burning much of the city in the process. This incident was blamed on palpable human error. The terrain around Malmedy was quite different from that near Zulpich, thirty-three miles away, and misidentification in good visibility seemed impossible.

On Christmas Day itself, four B–26s revisited Malmedy and dropped a further sixty-four 250-pounder bombs into the ruins. Aware that they had missed their true target, St. Vith, the pilots were apparently satisfied that they had hit an alternative target in Born. Again visibility was excellent, and once again the blue-on-blue was attributed to human error.

So frequent were these "human errors" that many American ground troops opened fire on their own planes as soon as they saw them. A staff officer of the First Division reasonably opined on July 7, 1944: "I wish you would tell the Air Corps we don't want them over here. Have them get out in front and let them take pictures, but no strafing or bombing."

The German offensive in the Ardennes in the winter of 1944, popularly known as the Battle of the Bulge, saw some of the most confused fighting of modern times. In the appalling conditions—low cloud, mists, and snow blizzards—friendly fire was common, much of it provided by American aircraft. On December 24, 1944, however, flying conditions were excellent with good visibility, yet this did not prevent a series of friendly fire incidents. A squadron of P–38s attacked the village of Buisonville, recently occupied by units of the U.S. Second Armored Division, and killed an American officer and wounded another. At Bastogne, P–47s strafed and bombed troops of the U.S. 101st Airborne Division.

In the fighting between the Salm and Ourthe rivers, units from the American Third Armored Division were heavily hit by friendly fighter bombers. On Christmas Day, 1944, near

Grandménil, American tanks were attacked by a flight of eleven P–38s from the 430th Fighter Squadron, which mistook them for Germans. In the carnage that followed, the Americans lost thirty-nine men killed and over a hundred wounded. The responsibility for this blunder lay not so much with the pilots as with the neighboring American Seventh Armored Division, which had called in the air strike. In spite of the fact that the tanks of Third Armored were carrying orange identification panels, the American pilots were not deterred and continued to attack them. Clearly, coordination between divisions and their allocated air supports was highly suspect. Human error, whether bureaucratic or military, was at the root of the problem, and yet one wonders why individual pilots could not have responded with the individual initiative that has increasingly become a demand of modern warfare.

America's Undeclared War With Switzerland

On September 5, 1927, George and Ira Gershwin's new musical, *Strike Up the Band*, opened in Philadelphia. Based on a script by George S. Kaufman, the show was a satirical comedy that included the most unlikely situation that even the most ardent of American patriots could contemplate—a war with Switzerland—and in defense of Fletcher's American cheese, at that. In 1927, it was all good, harmless fun, of course, and in the show the Swiss seemed to take that war as an opportunity for a display of Swiss hospitality. But on April 1, 1944, the Swiss were feeling far from hospitable when Kaufman's fantasy became reality and American bombs rained death and destruction on the undefended Swiss city of Schaffhausen. The undeclared Swiss-American war of 1944–45 had begun. Soon Swiss fighter planes were shooting down American bombers, killing American aircrew, while Swiss towns suffered a number of damaging raids in which many civilians died. And in spite of the best efforts of politicians and diplomats on both sides of the Atlantic, the killing went on, unintentionally on the part of the Americans, but in grim earnest by the Swiss, who were determined to defend their territory and their rights as a neutral state. This fantasy war was one of the strangest manifestations of friendly fire and one imbued with the most serious consequences.

Allied bombing of southern Germany after 1943 was fraught with great difficulty. Bad weather over the mountainous terrain on the Swiss-German border, combined with faulty navigation and pilot error, produced a series of friendly fire incidents in which American and—to a much lesser extent—British aircraft invaded Swiss airspace and dropped their bombs by mistake on Swiss targets. In such cases it was hardly adequate to inform the Swiss authorities that a degree of accidental damage or friendly casualties was to be expected in modern warfare. To the neutral Swiss, such attacks were acts of war and would provoke immediate retaliation. The Americans were politely informed that single aircraft violating Swiss airspace would be escorted to a landing field and forced to land there, while formations of two or more bombers would be attacked without warning. Soon after the Schaffhausen raid, Swiss aircraft, acting on this policy, shot down an American bomber and impounded another, after forcing it to land at Dübendorf. The Swiss hoped that they had made their point. Yet friendly fire incidents increased rather than diminished after this display of force. To the embarrassment of the American authorities, their air commanders could find no foolproof system to stop their crews overflying Switzerland. When Ambassador Leland Harrison apologized to the Swiss for the Schaffhausen raid and offered to pay compensation, he was presented with evidence of what the Swiss foreign minister described as the "deliberate attack" by fifty American planes on Schaffhausen, causing casualties of over a hundred dead and wounded.

At first, the Americans were unwilling to accept all the blame, pointing out that there had been earlier incursions of Swiss airspace by British planes. But when Carl A. Spaatz, commander of the U.S. Strategic Air Force in Europe, collated all the evidence, it was obvious where the fault lay. On April 1, 1944, two American bomber groups had indeed flown over Schaffhausen by mistake, though their crews were insistent that their bombs had missed the town. Spaatz duly apologized to the Swiss, though he frankly felt he had more important things to worry about, like the D-Day landings in Normandy, only a few weeks away. He knew that bad weather and heavy cloud cover was a common feature over Switzerland and southern Germany at that time of year and felt that it was all much ado about nothing. But when Spaatz's views were leaked to the press, there was an outcry in Switzerland, intensified by meteorological evidence that showed that weather conditions over Schaffhausen on April 1 had been excellent, with near perfect visibility. In Germany, the Nazi propaganda machine accused the Americans of "war crimes." The Americans were faced with a public relations disaster. Only generous compensation combined with measures to ensure that there was no repeat of such friendly bombing would satisfy the Swiss.

An investigation of the Schaffhausen raid revealed that adverse conditions over France, with heavy cloud and strong winds, had broken up the bomber formations, scattering the planes over a wide area. Their original target had been the German town of Ludwigshaven-am-Rhein, but when gaps in the cloud enabled the Americans to identify a city on the east bank of the Rhine, they did not realize that they had been blown across the Swiss border and were, in fact, fifty miles from their target, flying over Schaffhausen. At this stage, there was still a chance that the tragedy could have been avoided. Standard American procedure precluded bombing a target within fifty miles of Germany's borders unless a positive identification could be made. Two important targets—the benzol storage plant and the butadiene factory—should have shown the pilots whether they were over Ludwigshaven. Yet with only snatches of clear sky, none of the bombardiers could possibly have achieved such a positive identification, and so no bombs should have been dropped. On the other hand, human nature being what it is, it was asking a lot of a pilot to fly back across occupied France with a full bombload, particularly after having risked so much already. As a result, through human error, the target was not identified and American bombs were dropped on a city in neutral Switzerland, causing over a hundred civilian casualties. As a "one-off," the Schaffhausen raid would have been bad enough, but even as American officials tried to smooth the ruffled Swiss feathers, news of more incidents became public. The Swiss airforce fought fire with fire, and on April 13, its fighters shot down a damaged American bomber, killing the six crewmen. Now it was the turn of the Americans to protest.

In the space of three days in July 1944, a total of twenty-three American bombers were forced to land by Swiss fighters. The situation was getting out of hand. Yet, however much the politicians might fume at the diplomatic embarrassment, in the skies over southern Germany, American pilots were facing the choice of flying their bombers, sometimes damaged by flak or short of fuel, across France and the English Channel to their bases in England, or taking the easier and safer option of seeking internment in neutral Switzerland. Many chose the latter option, even jettisoning their bombs on Swiss territory rather than attempting a crash landing loaded with high explosives. On July 19, the castle of Weyden, home of the president of the International Red Cross, was struck a direct hit by a damaged U.S. bomber, abandoned by its crew.

Although these repeated incidents of friendly fire were viewed in an extremely serious light by U.S. politicians and diplomats, not everyone in the armed forces shared their concern. Senior members of SHAEF—both British and American—were convinced that the Swiss general staff was full of German sympathizers. Moreover, the fact that Switzerland—quite legally as a neutral state—had continued to trade with Nazi Germany, convinced some commanders that the Swiss deserved everything that was coming to them. Yet this was no answer to the most important question that needed to be asked: Why were so many American planes flying off course and bombing the wrong targets? One optimistic—and specious—theory was that German pilots were flying captured and repaired American planes over Switzerland in an attempt to score a sensational propaganda victory. But this was grasping at straws. There were quite enough verifiable incidents of American friendly fire to render any German cooperation quite unnecessary. On October 29, 1944, American bombers attacked the railway junction at Noirmont, even though Swiss flags were painted on a number of the village roofs and were quite unmissable in the good visibility that day.

On December 25, bombers from the U.S. First Tactical Air Force bombed the town of Thayngen in Switzerland, on the mistaken assumption that they were attacking the Singen railway bridge in Germany. The American pilots blamed cloud cover, but again it was obvious that a serious error in identification had been made, in which case the bombs should not have been dropped. In the early days of 1945, the Swiss were in the front line again, as U.S. bombers attacked Chiasso and the hydroelectric plant at Brúsio in the Puschlav valley. On February 22, President Roosevelt sent a special representative, Laughlan Currie, to Switzerland to apologize for the serious and continuing violations of Swiss airspace. Currie went to Schaffhausen to lay a wreath on the graves of the civilians killed by American bombs on April 1, 1944, but his timing could hardly have been worse. As he was honoring the dead in Schaffhausen, American bombers made their most widespread attacks on Swiss territory. Thirteen separate incidents occurred, and at Stein-am-Rhein—just twelve miles from where Currie was making his gesture—seven civilians were killed and some sixteen injured. This time it was impossible to blame adverse weather conditions, for conditions were bright and sunny. Currie's mission was fatally undermined. All he could do was to offer compensation.

At SHAEF, General Eisenhower was smarting from the criticism he was receiving from all quarters. He knew that the percentage of friendly fire incidents was minute compared to the thousands of successful missions carried out by British and American planes over Nazi-occupied France and Germany. Yet it was the mistakes that were getting all the headlines. When, in early March, six U.S. B–24s dropped twelve tons of explosives on Zurich, while others hit Basel with more than sixteen tons, the reputation of the American air force seemed on the line. In Washington, General Marshall ordered Ike to send Spaatz to Switzerland to see if he could do a better job of explaining why his bombers could not fly straight or bomb the right targets. At a crucial stage in the war, it was an incredible decision to uproot Spaatz from his London office and send him to the small French town of Annemasse, dressed in civilian clothes and a Tyrolean hat, to meet the Swiss top brass. It could have been a scene from *Strike Up the Band.* But the Swiss did not appreciate the honor that was being paid them. Perhaps Spaatz should have come barefoot and dressed in a sheet. They presented him with a full list of grievances, and it was a pretty long list at that. Their demand for compensation was readily accepted by the American delegation, and Spaatz expressed his readiness to prohibit bombing within 150 miles of the Swiss border unless precise identification was possible. Within fifty miles of Swiss territory, no bombing of any sort would take place. Had these restrictions been applied a year earlier, they could have had a significantly weakening effect on the Allied bombing campaign against Germany. At last, it seemed that the problems caused by friendly fire were coming to an end. The compensation finally paid to Switzerland for bomb damage amounted to some $18 million, settled in October 1949. Yet no one ever really solved the problem of why so many American pilots mistook their targets over so long a period. The bombing of Zurich in March 1945 was particularly puzzling. Situated as it was on a substantial body of water—the lake of Zurich—the city was clearly identifiable and was most unlikely to have been confused with any nearby German target. Moreover, to attack Zurich, the American bombers needed to have traveled deep into Swiss territory, and no sort of navigational error or adverse weather conditions seemed to provide an adequate excuse for bombing the city by mistake. In fact, the Zurich fiasco gave rise to one of the few occasions where those responsible for friendly fire incidents were both named and subjected to court-martial. On June 1, 1945, the pilot and navigator of the lead aircraft in the squadron that bombed Zurich were tried by court-martial at Horsham St. Faith in England. As it happened, the presiding officer on that occasion was none other than Hollywood star James Stewart, then a U.S. colonel. Both men were acquitted on a charge that they "wrongfully and negligently caused bombs to be dropped in friendly territory." In their favor, it was shown that their plane had been subject to equipment malfunction as well as poor visibility. Nevertheless, the final decision to drop the bombs was taken by the pilot, and the responsibility for the civilian casualties was his. As in most friendly fire incidents, it was human error that played the most significant part in America's unfortunate "war" against Switzerland.

The War in the Pacific

Although air amicide was much less of a feature in the Pacific than in the European theater of operations, it must not be regarded as insignificant. Distinguishing friendly forces from enemy troops was notably difficult in the jungle terrain of New Guinea, where there were numerous accidental bombings of friendly forces. According to Charles Shrader, the American troops involved in the recapture of Guam in July 1944 were constantly subject to friendly fire incidents. On July 21, the first day of the assault, members of the Twenty-second Marine Regiment on Agat were attacked by American aircraft. Three days later, units of the marines trying to break out of their bridgehead called in air support from the navy's planes. When it came, it proved to be more damaging to them than to the Japanese. The problem was that the two battle lines were far too close to allow the navy planes to drop their bombs with any confidence of avoiding friendly troops. The outcome was that seventeen marines died or were wounded in the attack. On August 4, the command post of the Third Battalion of the Twenty-first Marines was demolished by bombs from two B–25s. Not content with this, the two planes then strafed American personnel in the area. While the marines were suffering these blue-on-blues, the army also had its hands full fending off American attacks. On July 28, a company of the 305th was bombed and strafed by U.S. planes just after it had fought its way to the summit of Mount Tenjo, on southern Guam. This company only avoided destruction when one of its number, risking his life in the process, managed to spread out an identifying panel visible from the air. Further bombing and strafing incidents occurred to units of the 305th and 307th near Mount Santa Rosa and Yigo.

So frequent were the air attacks on friendly troops that the ground forces eventually lost all confidence when U.S. aircraft were called in to support them. Shrader reports the following exchange between General Krueger and General Kenney, which seems to sum up the feelings of infantrymen everywhere during the Second World War: "Ground-pounders" were "trigger-happy."

> I must insist that you take effective measures to stop the bombing and strafing of our ground forces by friendly planes . . . These repeated occurrences are causing ground troops to lose confidence in air support and are adversely affecting morale.[48]

KOREA AND VIETNAM

A tragic blue-on-blue occurred during the Korean War to British troops from the Argyll and Sutherland Highlanders. On September 23, 1950, the Highlanders had just captured Hill 282 from the Communist North Koreans and had called in air support against nearby enemy troop concentrations. They identified their own positions with recognition panels visible from the air but when the American Mustangs arrived, they ignored the signals and plastered the British position with napalm. Seventeen members of the Argylls were killed and seventy-six wounded. Furthermore, they were forced to evacuate the hill that they had fought so long and hard to capture. Hill 282 was reoccupied by the Communists, but Major Kenneth Muir and his thirty surviving men promptly fought their way back up the hill and retook it, Muir dying in the moment of victory. For his courage and leadership, Muir was awarded a posthumous Victoria Cross.

The problems of close ground support by the U.S. Air Force that had produced so many friendly casualties in World War II were even greater in Vietnam as a result of the acutely hostile environment. The planes were faster and more deadly in performance, and, in their ground support role, they were assisted by attack helicopters, yet the mainly jungle terrain of the battle zones made blue-on-blue incidents even more likely than in previous wars. The increased speed of their aircraft and the need for almost instant responses to a swiftly changing battle situation made pilot errors even more costly. Added to the problems of identifying friendly forces was the factor of mechanical malfunction by the increasingly complex hardware. Helicopters were particularly prone to technical failure.

On March 3, 1968, near the village of Go Vap, a blue-on-blue incident resulted from a helicopter hitting an air pocket, causing the nose of the plane to dip violently. Unfortunately, this unplanned maneuver occurred just as the crew was firing two rockets in support of an American platoon from the Ninth Infantry Regiment, which was caught in a Communist ambush. Just three Americans were wounded in the incident, but the way in which it occurred was disconcerting for ground troops entirely dependent on the usually reliable UH–1 helicopters for ground support.

In the confused fighting of a jungle battlefield, identification of friendly troops was the greatest difficulty for the jet and helicopter pilots in Vietnam. Friendly and enemy troops could and frequently did occupy almost exactly the same grid references, requiring pinpoint accuracy in bombing runs, something that was often beyond the capability of the pilots. Both planes and pilots were being asked to perform beyond their technological—and their human—limitations.

In one case, cited by Charles Shrader, the ground troops—members of the Vietnamese Civilian Irregular Defense Group—marked their position in the thick jungle by using green smoke. Two U.S. B–57s were called in to give ground support, but one strafed the designated target and managed to hit friendly troops, killing four men and wounding twenty-eight. The reason for this error was not simply the poor visibility afforded by the jungle but the fact that prior to his attack, the pilot was given numerous changes of target area and bearing. These new data clearly confused him and led to human error. Thus, even though the pilot had the assistance of an airborne forward air controller, it was still possible for him to be overloaded with data to the point where he, rather than the technology he controlled, malfunctioned.

The involvement of forward air controllers in ground support should have decreased the incidence of friendly fire in modern wars. However, the introduction of another fallible human element into the decision-making process of air-to-ground operations, though essential in view of the technical demands of flying and fighting a state-of-the-art combat plane, had added another factor to an already complex chain. Poor coordination between air controller and pilot and the communication of flawed data were at the root of a large number of blue-on-blue incidents in Vietnam and more recently in the Gulf. Literally, the more minds that are involved in taking a decision about where and when to drop bombs, fire rockets, or strafe ground troops, the more chances there are for human error. Shrader cites an example from Vietnam in 1968. An FAC controlling the mission of an F–4D aircraft armed with an M–117 bomb, marked a target just over two hundred yards from friendly troops. It was too tight for marking, particularly as the friendlies had put up no smoke for guidance and the FAC's guidance rocket went seventy-five yards west of the intended target. The pilot now made two errors of his own, to compound the FAC's inaccurate rocket guide. He misinterpreted the position of the friendly troops and incorrectly estimated the position of the target, which lay between the friendlies and the inaccurate rocket guide. The result was that the bomb landed squarely in the middle of the U.S. troops, killing three men and wounding twelve others.

The performance of the helicopter gunship, first used in Vietnam and then in every U.S. operation since, has revolutionized ground-support operations. However, Vietnam was in some ways the worst environment for such a weapon to be employed effectively. The jungle terrain made observation almost impossible and shrouded enemy antiaircraft guns. Nevertheless, the helicopter offered such advantages in mobility and the increased accuracy of its firepower that its success was assured in spite of the occasional blue-on-blue incidents that were a consequence of its unusual design characteristics. As Shrader has pointed out, the helicopter was as capable of misidentifying targets as the fixed-wing aircraft and its pilots just as liable to human error, yet its rotary blade and its proneness to instability in high winds carried with it particular problems for its pilots and the ground troops it was supporting. On August 27, 1967, a CH–47 helicopter was supporting units from the Twelth Infantry, who were engaged with enemy forces. As the helicopter moved over a company of the Second Battalion, its door gunner was shot and killed, by enemy fire. But the gunner was in the act of firing as he was killed and his grip on the trigger did not relax, so that he sprayed bullets into the friendly troops below. This unfortunate incident was a consequence of the nature of the helicopter as an aerial gun platform and could hardly have occurred in the case of any other aerial weapon. Even more unusual was the incident involving a fractured traverse rod on a door gun, which caused the helicopter gunner to fire into his own cockpit, wounding the pilot.

One element of friendly fire in the air that has always remained the closest of closely guarded secrets is the degree to which human irresponsibility has played a part in blue-on-blue. Foot soldiers since 1939, perhaps even earlier, have not always been convinced that their airman colleagues have been as careful as they could be when it came to supporting the infantry. The all-embracing category of human error has usually been enough to deter the investigator from pursuing individual motivation when it came to friendly fire. Yet what evidence there is does tend to suggest that some aerial blue-on-blues have resulted from drug- or drink-related carelessness on the part of pilots and aircrew, or at least that overexuberance and youthful high spirits have occasionally played their part. Shrader suggests that two examples from Vietnam that resulted in a number of deaths may have been caused in these ways. During August 1969, near the town of Pleiku, a UH–1H helicopter crewed by inexperienced personnel on their first flight responded to a command from the crew chief to fire on some smoke rising above the trees. The helicopter went directly into action without any attempt at identifying the target and fired straight into an American unit. In 1971, another UH–1H helicopter was shot down by American infantry from Fire Support Base Mary Ann, near Chu Lai. According to Shrader, indiscipline was the major cause of this disaster, in which the ground troops may have engaged the helicopter "for a lark." Such irresponsibility is an aspect of warfare that reflects badly on everyone, yet is an almost inevitable consequence of human involvement in warfare. A percentage—however small—of the military personnel of any army will be "gung ho," or "trigger-happy"; will love killing for its own sake; will commit atrocities, kill prisoners, rape women, torture suspects, and "frag" their own officers; will take drugs or be drunk on duty; will run away when the first shot is fired; and so on. However well-drilled and professional an army, the stress of operational conditions will bring out the best in some men and the worst in others. While human beings engage in warfare, they will always face two enemies: the enemy who points the guns and rockets their way, and the enemy within, the self that they did not know they possessed until their life was placed in the balance day after day.

GRENADA, 1983

During the American invasion of Grenada in October 1983, air strikes by Corsair IIs were used against Cuban and Grenadian People's Republican Army positions east of the town of St. George's. Unfortunately the fighter-bombers attacked Fort Matthew, an old British colonial fortress that was being used as a mental asylum. As bombs and rockets exploded in and around the fort, dozens of pathetic inmates ran screaming into the streets of the town. Although the fort was clearly marked as a hospital on the American maps, the pilots had confused it with other hill fortifications nearby. The consequences were that the Corsairs killed twenty-one helpless people and wounded hundreds of others. It was a shocking and thoroughly discreditable incident.

At 1600 hours on October 27, a scout platoon from the Third Battalion of the 325th Regiment was brought under sniper fire in the south of the island. Shots also passed over a jeep carrying the Air Naval Gunfire Liaison Company, whose task was to coordinate naval and air support for ground troops, notably the Third Battalion of the 325th. A chief warrant officer in the jeep spotted the house from which he thought the firing was coming and decided to call for support. Unfortunately he was not in contact with the Third Battalion at that time, and so on his own initiative, he called in a Spectre gunship, only to be told that none was available. He then made contact with the flight leader of four A–7s and gave the target details, which were a white house with a red roof, on a ridge north of a drive-in movie. He gave his contact pilot a bearing of 270 degrees from the sugar mill.

The A–7s made three passes over the target area, and the

chief warrant officer was satisfied that they had located the target. But when the A–7s began their bombing run, it seemed to the officer in the jeep that they were coming in on the wrong heading. The lead pilot was heard to say on his intercom that he could see people near the house. The ground officer knew this was wrong and called the pilots to abort the mission, but it was already too late. As he spoke, the leading aircraft opened fire on the wrong building, which turned out to be Colonel Stephen Silvasy's Second Brigade Tactical Operations Center (TOC), employed in coordinating the ground operation. The A–7 pilots had fired not into a white building with a red roof, north of the drive-in movie, but into a gray building to the west of the drive-in. In the operations center, seventeen men were wounded, including Sergeant Sean Luketina, who later died of gangrene after his legs were crushed. So out of touch was the chief warrant officer on the ground that he called for the A–7s to make another run, and when they refused and returned to the carrier *Independence*, he called in A–6 Intruders, which dropped bombs—fortunately duds—around the stricken operations center. This was too much for Lieutenant Colonel George Crocker and his men of the First of the 505th, who were in that area, and they contacted their air liaison officer to cancel any further air support.

Death in battle is never a matter for levity, yet the manner in which a man leaves this world can provoke a wide range of emotions. If one laughs, it is not with disrespect for those who perished, but as a reaction to the bizarre chances that may conclude a life of serious and earnest striving. Therefore, I am concluding this section on aerial amicide with two of the strangest incidents of friendly fire taken from the whole history of the Second World War. In the first incident, two Indian soldiers died when they were hit on the head by tins of boiled beef parachuted from a British aircraft. In the second, an American plane, in an attempt to lighten its load while struggling to reach its home base, jettisoned its cargo of bags of rice, one of which landed on a jeep, killing two of the four occupants.

4

Blue-on-Blue in Naval Warfare

There have been notably fewer examples of naval friendly fire than of aerial or ground friendly fire. This may seem surprising in view of the long tradition of naval warfare, going back almost as far as war on the ground, yet the factors that have applied to ground troops are less relevant to warfare at sea. Certainly there is less room for individual human error, unless the individual in question is the commander of a vessel or fleet. There have been accidents in naval warfare, and examples of friendly forces firing on one another in poor visibility. Yet it has not been easy to reconstruct accurate examples of naval amicide where evidence is at best sketchy. The chaos of the galley battles of the ancient world offer few opportunities for observing amicide in a naval context. With ramming the most decisive weapon available to naval commanders, followed by boarding, it would have required an unusual degree of objectivity on the part of onlookers to feel free enough or safe enough to report on the origin of ship damage. Accidental ramming in great galley battles like Salamis or Actium would have passed unnoticed, as would stray arrows or wayward sword strokes. In fact, we do know of one example of friendly fire from the Battle of Actium, but it was recorded only insofar as the historian wished to discredit one of the leading Greek traitors of the day, Queen Artemesia of Helicarnassus, who was fighting with the Persians against the Greeks. Finding herself hemmed in by Greek galleys and fearing capture or destruction, the lady made her escape by ramming neighboring Persian vessels to force her way into open water. Even as late as the sixteenth century, naval warfare in the Mediterranean was still dominated by galley warfare, although naval guns were beginning to make a tentative debut. At Lepanto in 1571, the interlocking of Christian and Turkish galleys allowed Spanish and Italian pikemen and musketeers a firm base from which to clear the decks of the Turkish ships with superior infantry tactics.

Before the mid-nineteenth century, ships fired mainly by broadside, and failures by individual gunners, for example, are unlikely to be recorded by ships' captains or commanders. Certainly in the great naval mêlées of the age of sail, notably the enormous battles of the Anglo-Dutch wars of the seventeenth century, where groups of friendly ships often attacked individual enemy vessels, there cannot fail to have been a high proportion of damage inflicted on friendly vessels by stray shots. During the battles of the eighteenth century and the Revolutionary-Napoleonic periods, where the riggings of men-of-war were swarming with sharpshooters and the decks below sometimes crowded with marines from both friendly and enemy vessels, a proportion of casualties would have been caused by friendly fire. However, it has only been in the last hundred years or so that the evidence has become available to reveal instances of naval friendly fire.

As we have seen in so many examples from air and ground amicide, human error is the single greatest factor in friendly fire incidents. The failures of individual pilots can have devastating effects because of the nature of the weapons that modern planes can deliver. Individual infantrymen, from the American Civil War to Vietnam, are still limited in the degree rather than the frequency of their contribution to friendly fire. And so it is with the individual sailor at sea, unless by his rank he is able by his error to order ships or fleets to commit amicide.

After the extraordinary success of the Austrian Admiral Tegethoff against the Italian fleet at Lissa in 1866, during which his flagship had succeeded in ramming and sinking the Italian flagship, naval designers made a prominent bow ram a feature of all capital ships for half a century. In Britain, the ram as a weapon took on the same heroic mystique that had always been associated with the bayonet. As the ranges at which capital ships could engage grew greater with every development in naval gunnery, the ram represented an archaic weapons system—heroic but irrelevant. Even the doyen of modern battleships—the *Dreadnought* itself—was equipped with a ram and achieved its only success in battle not through its revolutionary twelve-inch guns but with its ram, when it rammed and sank a German U-boat in 1916. Yet during the late Victorian period—the heyday of British naval preeminence—British admirals still dreamed of coming to grips with the enemy in true Nelsonian fashion, first ramming the enemy vessel and then boarding it. Even as late as 1854, Sir Charles Napier, commander of the British Baltic Fleet during the Crimean War, reminded his men to sharpen their cutlasses before going into battle, while in the 1870s—at a time when naval guns had a range of ten thousand yards—Admiral Sir John Commerell criticized his chief engineer for not wearing his sword. When the man replied that there was not enough room in the engine room for swords, the admiral was aghast; "What do you mean? If the enemy has the good fortune to overpower us all on deck, how will you kill him if he comes down here if you have no sword?" For British admirals of the time, rams were an article of faith, ship-killers that were far more deadly than guns could ever be. Yet no British admiral was to enjoy the experience of a major fleet action between 1827 and 1914. How could the rams find ships to kill? The answer, in two famous cases, was by committing naval amicide—by killing their own kind.

Our first example of a naval blue-on-blue is one in which a simple arithmetical error made on a quiet, sunny day in peacetime unleashes terrible forces of death and destruction. The fact that the error was made by a man who, in the eyes of his subordinates, was incapable of error, only served to make the incident more tragic and more deadly.

Vice Admiral Sir George Tryon, a tall, massive-chested, black-bearded colossus of a man, had a mind that was constantly at work trying to find ways to perfect the performance of the ships and men of Her Majesty's Mediterranean Fleet. One of his contemporaries wrote of him that Tryon conducted his work so skillfully as to prove every admiral arrayed against him his inferior. But this was just the problem. In 1891, nobody in the Royal Navy, however senior he might be, dared to question the orders of a man like Tryon. To do so would be to risk public humiliation.

Admiral Tryon had not needed wealth or family connections to start his career, as did so many senior officers in both the army and navy in Victorian Britain. He had risen to the top through sheer ability. His intellect allowed him to dominate everyone around him and, as another officer wrote: "Most people felt no use arguing with George Tryon, and that it was better to acquiesce quietly." But this very quality was to bring ruin on him. Tryon did not want obsequious subordinates; he was looking for men to show drive and initiative in the way that he always had. Instead, he found the majority of officers in the navy were afraid of his reputation, so that none dared to stand up to him or ever suggest that he could be wrong. Most of the officers who served with him simply wanted him to tell them what to do. So while Tryon tried to test them and develop their own skills, they were afraid to show how inadequate they were. Tryon's greatest problem, of which he was only partly aware, was that the officer who most felt this way was none other than his second in command, Rear Admiral Sir Hastings Markham, a man who has been described as "anxious, conforming, hidebound, conventionalist, dedicated to staying out of trouble and not displeasing his superiors." In a war situation, this match-up would probably have proved fatal to one or the other, but during peacetime, it should have been possible for the unhealthy relationship to continue until one or other of the senior men was put out to pasture. But Tryon was not content to live peacefully in peacetime. For him, periods of peace were merely rests between wars, and should be used to hone to perfection the mighty weapon that their lordships of the admiralty had left in his charge.

On taking up command of the Mediterranean Fleet—the strongest naval weapon in the world in 1891—Admiral Tryon set about introducing new ideas and challenging existing systems. Inevitably, he met not opposition but inertia. Accepted methods had been comfortable; Tryon was never comfortable. Sir Geoffrey Phipps Hornby observed that under Tryon there was none of the friendly banter that used to characterize the

Mediterranean Fleet in years gone by. Part of the problem was that Tryon was attempting to improve the speed and efficiency with which his ships maneuvered, trying to obviate the need for cumbersome flag signals. His own "TA" system involved less a system of signaling than of simply "follow my leader." As Tryon wrote: "I have been long impressed with the importance of exercising a fleet from the point where the drill books leave off . . . It is apparent to me that a fleet that can be rapidly maneuvered without having to wait for a series of signal repetitions will be at a great advantage." But this was deeply worrying for his subordinates, who would now have to maneuver their ships without signals from the flagship. All Admiral Tryon would do would be to raise the signal "TA," and the rest of the fleet would then conform with the movements of the flagship. What would happen if, in action, the flagship was disabled, Tryon does not seem to have considered. Conservatives at the admiralty were outraged, and the *Times* declared that the new system was "Unsound in theory and perilous in practice." But, although they sniped at him from a distance, nobody dared to tell Tryon that he was wrong. Prominent among Tryon's secret critics was Rear Admiral Markham, who lacked the courage to admit as much to his face. In fact, Markham's thirteen months serving with Tryon had been little less than a nightmare. On maneuvers, Markham simply could not deal with the constant tests set by Tryon to keep the fleet on its toes. And if Markham refused to express doubts to his superior, Tryon was quick to criticize his second in command in public. Tryon was not a man to suffer fools gladly, and in his view, Markham was a fool.

On June 22, 1893, the eleven ironclad battleships of the Mediterranean Fleet left harbor at Beirut and put to sea on maneuvers. The sea was as calm as a millpond, and the heat was oppressive. Tryon was feeling lethargic after a heavy lunch, and the officers and men aboard the flagship *Victoria* went about their duties as if the five days' leave they had spent in the souks of Beirut needed to be cleared from their systems by a short, sharp shock. The *Victoria* was a brand-new battleship, just three months old. Its massive guns—which supposedly made it the most powerful ship afloat—were feared more by its own crew than by the enemy. When they fired its main armament, the blast often buckled the deck and damaged the bridge, as well as playing havoc with the paintwork. In many respects, the *Victoria* and its sistership, the *Sans Pareil*, were unloved, especially by their own crews.

It is safe to surmise that none of this was passing through Admiral Tryon's mind that warm summer's afternoon off Beirut. Instead, he was thinking of setting the captains and crews of his fleet a really difficult test. He turned to staff commander Thomas Hawkins-Smith, saying: "I shall form the fleet into columns of two divisions, six cables apart, and reverse the course by turning inwards." Hawkins-Smith immediately felt uneasy: Six cables was just twelve hundred yards, and if the two columns were to turn inward, that would surely be too close for comfort. Was this a deliberate mistake to see if he was paying attention? He plucked up courage and said: "It will require at least eight cables for that, sir." Tryon thought for a moment and then agreed. "Yes, it shall be eight cables." After a few moments, Tryon's flag lieutenant, Lord Gillford, came into his cabin, and Tryon told him to make the signal. "Form columns of divisions line ahead, columns disposed abeam to port. And make the columns six cables apart." To confirm the signal he handed Gillford a scrap of paper on which he had scribbled the single figure "six." The flag lieutenant left without questioning the order and set about preparing the signal. Within minutes, the flags were fluttering and had been acknowledged by the other battleships of the fleet. Looking up from his work in the forebridge, Hawkins-Smith experienced a moment's uncertainty—that was the wrong signal. He hurried over to Gillford. "Haven't you made a mistake? The admiral said the columns were to be eight cables." But Gillford showed him the scribbled figure on the paper to confirm that Tryon had specifically ordered six cables. Hawkins-Smith was in a quandary. On the strength of his earlier conversation with the admiral, he sent Gillford back to ask Tryon to confirm "six cables." Tryon was not pleased to be cross-examined by a subordinate and told Gillford brusquely: "Leave it at six cables."

Gillford was now thoroughly alarmed. A simple calculation told him that the combined turning circles of ships like the *Victoria* and the *Camperdown*—the flagship of Rear Admiral

Markham, leading the other column—was eight cables, or sixteen hundred yards. The only answer must be the unthinkable—Tryon must have made a mistake, confusing the radius of the turning circle with its diameter. The maneuver he had ordered was impossible. Captain Archibald Maurice Bourke, the *Victoria*'s commander, saw immediately that the signal was wrong. As the ship's captain, he rather than Admiral Tryon was responsible for its safety. Yet, as Bourke later commented, "open criticism of one's superior is not consonant with true discipline," and so he chose to do nothing but to grit his teeth and pray. Two of the most powerful battleships afloat, armed with ship-killing rams, had just been ordered to turn toward each other in such a way that a collision was inevitable. At least three men on the *Victoria* were quite aware of this, along with others presumably on board the *Camperdown* at the head of the parallel column of ships. In a military sense, each commander was holding a loaded pistol to the head of the other, and both were content to fire rather than admit that a mistake had been made. It was an extraordinary situation—amicide in slow motion and by common consent.

The great fleet was now traveling at about nine knots toward the coast of Syria, and a turn of some kind would have to take place soon before the battleships ran aground. Tryon was growing angry; why had the *Camperdown* not yet begun to turn? Markham was holding up the entire maneuver. Tryon ordered a signal made to *Camperdown*: "What are you waiting for?" It was a public rebuke, and Markham would have to obey, whatever the consequences. Markham was puzzled, but clearly presumed that Tryon intended his column to turn first. But he was wrong, and as the *Camperdown* began its turn, so did the *Victoria*. Trying to conceal his anxiety, Bourke remarked to the admiral: "We had better do something, sir, we shall be too close to [the *Camperdown*]." Tryon ignored him, absorbed in the awesome potential of his own flawed geometry. Bourke's voice began to rise slightly: "We are getting too close, sir! We must do something, sir! May I go astern?" Tryon's voice was no more than a whisper. "Yes, go astern." "Full speed astern both screws," bellowed Bourke, but he knew that it was far too late now—a collision was inevitable.

All Bourke could do now was to try to lessen the impact of

the collision by ordering the watertight doors to be closed. It might do something to save the ship. As the *Camperdown* headed toward the *Victoria*, for a few seconds Tryon and Markham were within hailing distance, and Tryon shouted through cupped hands: "Go astern—go astern." He should have saved his breath, for at that moment the *Camperdown*'s huge ram ripped nine feet into the *Victoria*'s side, forcing it seventy feet sideways and wounding it fatally.

As the two battleships staggered under the impact, a yeoman passed a message to Tryon from the *Camperdown*. It was Markham's reply to his signal, demanding to know why *Camperdown* had not begun its turn. It was simple: "Because I did not quite understand your signal." Less than fifty yards away, Markham and his bridge officers were gazing in fascinated horror at the damage they had just inflicted on the fleet flagship. Tryon bellowed across: "Go astern, go astern. Why didn't you . . . ?" There was really nothing more to say but they were not quite his last words. "It is all my fault," he was later heard to mutter.

While time seemed to stand still on the flagship, the other battleships of the fleet had begun lowering boats to rescue Tryon's crew, but the admiral would not give up his ship and angrily sent the boats back. It was his second mistake that day. He believed that the *Victoria* had not been struck in a vital spot; he was wrong. The flagship's low forecastle was already under water, and the sea was pouring into the ship through every open porthole and door. The *Victoria* heeled to starboard, and water poured in through the turret apertures. Within five minutes of impact, the bows had sunk some fifteen feet. Everything was happening too quickly, and many of his men were going to die because he had rejected the offers of help. In just four more minutes, the *Victoria* began to slip beneath the waves. The men in the boiler rooms, receiving no orders to abandon their stations or even stop the engines, were drowned to a man. One lucky survivor was John Jellicoe—then a lieutenant but later commander in chief of the Grand Fleet in the First World War—who had been in sick bay on the *Victoria* with Malta fever and had escaped in his pajamas.

Admiral Tryon made no attempt to save himself. We can only guess what he was thinking as his flagship sank and his whole world came crashing down about his ears. He must have realized that the fault was his alone, and he had no wish to outlive the disgrace. Hawkins-Smith was one of the last men to see him, standing "perfectly calm and collected to the last" and dying "as he had lived, a brave man."

With Admiral Tryon, 357 officers and men died as the result of an inexplicable disaster on a sea as calm as glass and in perfect visibility. Yet the disaster was not really inexplicable; it was the result of a simple error in calculation by a man who considered himself incapable of making it. What is more difficult to understand is why Tryon seemed unprepared to admit his error and correct it, thereby preventing a disaster and the deaths of so many of his men. The exchanges with his officers over the figures "six" and "eight" indicate that Tryon did not enjoy being reminded of his first mistake and preferred to risk the consequences of it, whatever those might be, rather than admit to younger and more junior men that their commander was not perfect. The only conclusion that can be drawn is that Tryon suffered from an ego of such staggering proportions that it could not admit even a semblance of doubt. There was a perpetual risk that the collapse of such a brittle persona, built on a platform of other men's fears and adulation, would be accompanied by disaster. That the disaster was of such proportions was both a misfortune and a disgrace. Modern psychiatric selection procedures might have warned against allowing a man of Tryon's character to control a powerful fleet, the lives of hundreds of men, and the destiny of an entire empire.

After holding a somber funeral ceremony at sea, Markham ordered the fleet to return to Malta. The court-martial—inevitable in such a case—reached the decision that Sir George Tryon had been responsible for the disaster but also regretted that Markham had not questioned the fatal order more effectively before beginning the turn that led to the collision. This did not go down well in military circles. The army commander in chief, the Duke of Cambridge, complained: "A good deal has been said of late as to freedom being given to inferiors to question and disobey the orders of a superior officer. Discipline must be the law, and must prevail. It is better to go wrong according to orders than to go wrong in opposition to orders." Markham never held another important command, and the admiralty made it clear that they felt the *Camperdown* had been ineptly handled during the crisis. But none of this helped to restore the reputation of Sir George Tryon. His one mistake had been enough to destroy a career of unparalleled achievement. Command is a lonely business, and nobody will ever know what was in Tryon's mind when he gave the order for the maneuver that killed him. We can only agree with a comment made by Admiral Kerr at the time: "Sir George Tryon was not a person who was agreeable on being asked questions or cross examined." The result of Tryon's unapproachability was one of history's most extraordinary naval blue-on-blues.

The ramming and sinking of a Victorian ironclad by one of its own kind had occurred on at least two occasions before Tryon's memorable mathematical inexactitude. But in the case of the sinking of the *Vanguard* by the *Iron Duke*, there were extenuating circumstances. Visibility was so poor on the day of the fatal collision that the commander of the *Vanguard* was hardly to blame. This was not the view of the British admiralty of the day, but under the circumstances it is difficult to judge either Lieutenant William Hathorn or Captain Richard Dawkins too harshly for what happened.

On July 29, 1875, the First Reserve Squadron of the Home Fleet, under the command of Vice Admiral Sir Walter Tarleton, left Portland for exercises off the coast of Ireland. Tarleton was flying his flag in the *Warrior*, Britain's first iron battleship, and with him were three other ironclad battleships, *Hector*, *Vanguard*, and *Iron Duke*, traveling at eight knots in two divisions. On board the *Vanguard*, leading the port division of Tarleton's squadron, Lieutenant William Hathorn had just relieved Captain Richard Dawkins, who had gone below to rest in his cabin, after a period on watch. There seemed to be no cause for concern at that time, and, coincidentally, at almost exactly the same moment, Captain Henry Hickley of the *Iron Duke*, second ship in the port column, also went below, leaving an officer of the watch on deck. Both captains—Dawkins and Hickley—had been on deck for many hours, and both were tired. Suddenly and unexpectedly, a thick fog descended, and visibility was rapidly reduced to less than the length of one of the battleships. As each of the ships was traveling at twice the rate that was normally permitted in fog, each would obviously have to reduce speed. But how were they to achieve this rapid speed reduction in safety? So sudden had been the onset of the fog that there had been no time for a signal to be sent from the admiral to reduce speed. Dawkins therefore returned to deck and discussed with Hathorn how to use the steam whistle to indicate his intention to reduce speed. They decided to sound the ship's pennants and reduced speed to six and then five knots. Dawkins was not sure of the correct signal and spent some time checking the signals book. But before he could do anything, he received an emergency report of a sailing ship "right ahead." This placed Dawkins in a dangerous position. If he continued on his current heading he would hit the sailing ship; but if he stopped, he would be rammed from behind by the *Iron Duke,* and if he swung away in either direction, he might easily collide with any of the other ironclads.

Dawkins decided that he had no alternative other than to stop and to swing slightly to port to allow the sailing ship to pass. He could hear Admiral Tarleton on the starboard bow sounding his pennants, but as far as he could tell, the *Iron Duke,* directly behind him, had made no signal at all. He placed lookouts all around the ship, but it was difficult to bring such a large vessel to a halt quickly. As the battleship slowed, the worst happened: The *Iron Duke,* apparently steaming at undiminished speed, drove straight into it from behind. In fact, the *Iron Duke* had turned slightly to port to avoid the possibility of running into the *Vanguard,* only to find that the *Vanguard* had done the same to avoid the sailing ship.

Captain Dawkins had only been on deck seven minutes, but in that time his ship had been fatally damaged, and he sounded his whistle continuously, calling for help. The *Iron Duke* quickly lowered its boats to help rescue the crew of the stricken ship. But was the *Vanguard* sinking? Chief Engineer Robert Brown reported that it was, in spite of closing the watertight doors, but Dawkins was not prepared to give in so easily. While his crew was taken off by the rescue craft, he struggled in vain to save his ship. He was eventually forced to leave—the last to go—leaving just one casualty behind, his pet dog.

A court-martial was held on September 10, 1875. Captain

Dawkins came in for heavy criticism for having chosen to save the crew rather than the ship. The court was entirely unsympathetic to the captain's dilemma and decided that he was traveling too fast, had left the deck at the wrong time, and had slowed down without correctly informing the *Iron Duke.* The court also believed that Dawkins should have gotten his pumps working and saved his vessel. The court concluded by blaming Dawkins for showing lack of judgment and neglect of duty in saving his men and losing his ship. The only redeeming feature, according to the admiralty, was that the *Iron Duke* had actually succeeded in doing what it had been built to do—sinking a battleship. After the court-martial, an anonymous ditty was composed that contained these pertinent comments:

Run into and sunk by another tin junk,
'Twas very good sport d'ye see.
"For it clearly showed what a fine ram she had got"
Said the Lords of the Admiraltee.

Just three years later, on May 31, 1878, an extraordinary disaster took place on a beautifully clear day off the Kent coast. On that Friday afternoon, a group of spectators stood on the cliffs watching a squadron of three German ironclad battleships, under the command of Admiral von Batsch, traveling slowly by in two columns. They had left Wilhelmshaven on May 6 and were on a training cruise to Plymouth. Von Batsch's flagship, the *König Wilhelm,* was leading the port division, and behind it came the *Preussen.* The starboard division consisted of just the single ironclad *Grosser Kürfurst.* As the German warships sailed by, the onlookers could see that two small sailing ships were crossing their bows, forcing the *Grosser Kürfurst* to swing to starboard to pass under their sterns. The *König Wilhelm,* in trying to avoid the same vessels, now made a terrible mistake. The inexperienced crew at its helm—a young petty officer and six raw recruits—got confused, and when ordered to "starboard the helm" did the precise opposite, causing the flagship to ram the *Grosser Kürfurst,* "ripping off the armour plates like the skin of an orange and sweeping away the quarter boats like strips of paper." Poor von Batsch was in his cabin when the disaster took place, and by the time he got on deck the stricken ironclad was sinking, with the loss of 284 men of its crew of nearly 500 in spite of the efforts of rescue craft from nearby Sandgate.

If sinking a friendly ship by ramming is the most obvious example of naval amicide, there have been other ways in which naval commanders and their vessels had damaged their friends or inflicted harm on their national interests. An early example of the latter occurred during the Crimean War and involved officers and ships of the British Baltic Fleet. The unprecedented depredations of one of Her Majesty's flag officers, Admiral Plumridge, commanding a flotilla of Her Majesty's ships, can only be regarded as an outrage in the context of the Royal Navy's worldwide reputation at that time.

On March 29, 1854—at the outbreak of the Crimean War—the British secretary of state for war, the Duke of Newcastle, had written to the admiralty explaining Britain's attitude toward the subject of friendly states and their neutral rights in the coming struggle:

> It is Her Majesty's anxious desire that the interests of humanity should be regarded in the war upon which Great Britain has now entered, to the fullest extent which its operations will admit; and I am to request your Lordships to give positive orders to Sir Charles Napier to respect private property wherever it can be spared, without a sacrifice of the objects of the war, and on no account to attack defenceless places and open towns . . ."[49]

Newcastle was aware that British operations in the Baltic might involve the Royal Navy in difficulties with neutral states such as Denmark and Sweden, as well as pro-British elements within Finland, then a part of the Russian empire. It was not in Britain's interests for its navy to be seen as damaging the property or interests of these states. Unfortunately the British commander in chief, Sir Charles Napier, seems to have signally failed to convey this message to all his commanders, notably in the case of Rear Admiral James Hanway Plumridge, commander of a squadron of paddle-steamers, who was ordered to reconnoiter the Aland Islands and blockade the coastal towns of Bothnia. During a five-week period, from May 5 to June 10, Plumridge in the *Leopard* carried out an "orgy of destruction" that prompted the king of Sweden to describe his actions as "barbarous and unworthy of our time." Plumridge and the "gung-ho" Captain Giffard of the *Leopard* undertook to "take, burn and destroy" in the style of a latter-day Viking. Had it been enemy property that he took, burned, and destroyed, there might have been some excuse for it. However, through a serious failure of intelligence, Plumridge waged economic warfare on Britain itself! On his orders, his men and ships committed amicide against Britain's allies as well as its neutral trading partners.

Failing at the outset to reconnoiter the Aland Islands as the fog was too thick and the floating ice too dangerous, Plumridge found a different outlet for his energies by seizing the vessels of the Finnish coastal farmers, plundering their equipment, and burning their sails. Leaning heavily on the unreliable Giffard for advice, Plumridge next raided the Finnish ports. In his journal, Giffard took an unseemly delight in listing the depredations of his British squadron, without considering for a moment that the people who were suffering were not Russians at all but Finns. Moreover, most of the products that his ships were destroying—notably timber and pine tar—were destined for Britain. Before the war, Britain had imported annually 56,500 barrels of tar from Finland—35 percent of Britain's total tar imports. Now he had devastated the vital tar trade. Nobody seems to have told Plumridge or Giffard that the local population was both pro-British and Britain's vital trading partners. It was often British produce—paid for in advance by British companies—that went up in flames to the delight of Giffard and his desperadoes. All that they were achieving by this friendly fire was to damage British trade, ruin simple farming folk, and drive the whole population into the arms of the Russians. Significantly, as time passed, the villagers began calling in Russian troops to help them defend their farms and villages against the marauding British squadron. At Uleaborg, Plumridge threatened a group of villagers, who were holding up a flag of truce, that he would bombard their church if there was any resistance. And when the Finns did fight back, Giffard described their action as "treachery."

Throughout the sordid episode, Sir Charles Napier was quite out of touch with the real damage that was being done. On June 18, he even proudly reported to the admiralty that Plumridge had "destroyed 46 vessels afloat and on the stocks . . . from 40,000 to 50,000 barrels of pitch and tar . . . a great number of stacks of timber, spars, plank and deal, sails, rope and various kinds of naval stores to the amount of from £300,000 to £400,000, without the loss of a single man." Their lordships were singularly unimpressed. In London, British merchants and trading houses, receiving news of the destruction of their property from their trading partners in Finland, demanded compensation from the government for these "unnecessary and barbaric" raids. The *Times* was quick to respond, condemning the British raids against a friendly trading partner. To make matters worse, the raids led to a wave of anti-British feeling in neutral Sweden. Admiral Plumridge was described as an arsonist and the British fleet as no better than Vikings. As one Finnish writer has observed: "These opening moves in the Crimean War stirred up real hatred of Britain, and at the same time caused loyalty to the [Russian] Emperor to become more pronounced. In their crisis the Finns looked to him for protection."

Plumridge's expedition was a catastrophic mistake. The rest of the British fleet knew that it would only make the Finns join the Russians in resisting them. On June 9, the villagers at Gamla Carleby, stiffened by a few Russian troops, drove off a British raiding party, killing fifty of the crew of the paddle-steamer *Vulture.* In London, Sir James Graham tried to distance the government from Plumridge's excesses, reminding parliament of the Duke of Newcastle's instruction, issued even before the fleet had sailed. Sir Charles Napier was held to blame for not controlling Plumridge and was relieved of his command. But the damage had been done, and Giffard and Plumridge had cost their country not only millions of pounds in compensation, but also the goodwill of neutral Baltic states and the support of the Finns in Britain's war with Russia.

Accidents, involving heavy loss of life, have been a feature of amicide on both land and sea, and in the air. The turret explosion on the battleship *Thunderer* on January 2, 1879, was one such disaster and, moreover, one that could have been avoided by a more progressive approach from the British admiralty. Even though breech-loading naval guns had been in use in some European navies as early as 1860, the British admiralty was slow in making the transition from the tried and tested muzzle-loaders that had served Britain so well since the time of Sir Francis Drake and the armada. There seemed little point in building bigger and better ships at great cost when their main armaments suffered the extraordinary drawback of having to be loaded externally. In addition, the shells were now far too large to be loaded manually, and so extraordinary methods had to be adopted to lift the massive weights into the barrels. A huge sixteen-inch gun, weighing over eighty tons, was built for HMS *Inflexible.* A shell from this monster could penetrate twenty-four inches of armor plating. The world's naval architects, notably the Italians, could only gasp with astonishment at the obstinacy of Britain's naval designers. Rather than equipping the *Inflexible* with breech-loaders for its huge revolving turret, the British had designed a deck housing to which, after each firing, the turret had to return so that it could have its barrels reloaded by a hydraulic system within the housing. So vulnerable was this device that had it been damaged in action, the great ship itself would have been totally disarmed and helpless. Enemy ships of far less strength than the *Inflexible* would have been able to close in to point-blank range and pound it with their smaller guns.

Eventually, it took a disaster aboard the battleship *Thunderer* to make the admiralty see the error of their ways. On January 2, 1879, the *Thunderer* was taking part in gunnery practice in the Gulf of Ismid, in the Mediterranean. One of her twelve-inch, thirty-eight-ton Woolwich muzzle-loading guns—small when compared with *Inflexible*'s huge sixteen-inch rifles—was being charged with eighty-five pounds of powder and a common shell when the shell exploded in a barrel of the ship's forward turret. The muzzle of the gun was blown off, causing terrible damage, in which two officers and nine men were killed, and thirty-five others seriously wounded by flying metal splinters. The accident was probably caused by double-loading the muzzle, but the committee of inquiry at first tried to claim that the shell had shifted after being hydraulically rammed home. Whatever the true cause, the accident was a clear pointer to the dangers inherent in clinging to the outworn habit of muzzle-loading. The breech-loading turret had been adopted in other navies, and Britain risked losing its naval supremacy by building ever larger models of an outdated system. The *Thunderer* disaster eventually convinced the admiralty that the time had at last come to accept that change was necessary. But the lesson in warship design had only been learned at the expense of a lot of self-imposed casualties.

Almost certainly the most remarkable example of amicide in naval history occurred in 1905 during the Russo-Japanese War, as the Russian Baltic Fleet sailed from the port of Libau in the Baltic to the Straits of Tsushima, off Formosa in the South China Sea. No voyage by any fleet in history has attracted such hostile attention from historians, and it is difficult to defend Admiral Rozhestvensky and his men, whose gross incompetence was at times farcical and at times tragic. The fleet appeared to suffer from a series of group complexes, being collectively neurotic, paranoid, and schizophrenic.

The Russo-Japanese War broke out in 1904 when, without any formal declaration of war, Japan made a preemptive strike against the Russian Far East Fleet. On the night of February 8, the Japanese torpedoed two Russian battleships and a cruiser in the harbor at Port Arthur. From that moment onward, Japan had maintained a tight grip on the seas that the Russians were never able to break. When they did try to break out of port in August, they suffered a resounding defeat at the Battle of the Yellow Sea. Yet, rather than convincing the Russian government that the war was lost, this setback only seemed to make them more determined, and they reached the decision that their Baltic Fleet should sail most of the way around the world to meet an enemy who had already defeated a powerful Russian naval squadron. To compound the difficulties, between the Gulf of Finland and Port Arthur there were no Russian bases at all, and the ports of neutral nations and even those of Russia's ally, France, would be closed to them. To supply the immense amounts of coal needed by the fleet—half a million tons at a conservative estimate—Admiral Rozhestvensky would have to rely on prearranged meetings at sea with sixty colliers of the German Hamburg-Amerika line, and the length of the journey would mean that his forty warships would need to recoal on as many as thirty occasions, each time in the open sea, subject to wild weather and heavy waves.

For all the impressive appearance of Rozhestvensky's quartet of battleships, the strength of the Russian fleet was more apparent than real. Its crews were of very low quality, few being drawn from the coastal, seafaring parts of Russia, and most of the sailors were simple, untutored peasants who received little training at sea, as the Baltic was iced over for half the year. With the greater technical demands of modern sea warfare, the lack of education of the Russian sailors was a particular drawback. During one training exercise, Rozhestvensky sprang a surprise alarm—"defense against torpedo attack." He waited on the bridge of his flagship, but nothing happened—no men took up their posts—for everyone, officers and men, was fast asleep. One British sailor described the Russians he met as "odorous, rough, coarse but a happy lot." An officer aboard the *Suvoroff* complained about his own gunners: "One half have to be taught everything because they know nothing; the other half because they have forgotten everything; but if they do remember anything, then it is obsolete." What nobody could have guessed at the outset was that some of the seamen were revolutionaries, who tried to foment unrest among the crews. It was not a happy situation for any commander.

As the fleet left the Russian Baltic port of Libau on October 16, the flagship set the pattern for the entire journey by running aground, and a cruiser lost its anchor, wasting hours trying to locate it. While this was going on, a destroyer rammed the battleship *Oslyaba* and had to return to Reval for repairs. But once these wrinkles had been smoothed out, the fleet moved peacefully into the narrow waters between Denmark and Sweden. Here reports that the Japanese had torpedo boats stationed along the Danish coast meant that the Russians were continually on the lookout for spies, saboteurs, or indeed Japanese warships disguised as trawlers or yachts. Chief Engineer Politovski wrote to his wife:

> Tonight there will be danger. We shall all sleep in our clothes and all guns will be loaded . . . We are afraid of striking Japa-

> nese mines in these waters. Perhaps there will be no mines; but considering that long ago Japanese officers went to Sweden and swore to destroy our fleet, we must be on our guard . . . Panic prevails aboard. Everyone examines the sea intently. The weather is glorious. The slightest suspicious-looking spot on the water is carefully watched. The guns are angled. The crew are standing about on deck . . . It is curious that we are so far from the theater of war and yet so much alarmed.[50]

This paranoia helps to explain the events of the next few days. The Russian government had paid its agents large sums of money to give advance warning of any Japanese surprise attacks. A certain Captain Hartling had been sent to organize counterespionage in Copenhagen, where he inhabited a fantasy world of spies and secret weapons, and from which he transmitted daily to the fleet. He reported that everyone, it seemed, was against them, and that the seas were awash with Japanese mines, submarines, and especially torpedo boats. This misleading and incorrect information led to mass hallucinations on the part of the Russian crews. Rozhestvensky ordered that "no vessel of any sort whatever must be allowed to get in amongst the fleet." When two fishermen were sent out by a Russian consular official to deliver a telegram to the flagship, they were nearly blown out of the water by the trigger-happy Russians. In fact, the telegram informed Rozhestvensky that he had been promoted to vice admiral by Tsar Nicholas II. Other terrors for the Russians were two silvery balloons seen in the distance but never traced afterward, which convinced the fleet that the Japanese were observing their every move.

Once in the North Sea, thick fog became a new enemy. An officer on the battleship *Oryol* wrote: "The world has become nothing but an infinite envelope of mist. Men looked like phantoms . . . The impression of mystery was intensified by the noise of sirens . . ." When the fog cleared, the fleet repair ship, the *Kamchatka*, was found to be missing. Suddenly, a signal was picked up by flagship. It was the *Kamchatka*, claiming that it was being chased by at least eight Japanese torpedo boats. Although it was a false alarm, it added to the air of tension.

Rozhestvensky now almost brought about a war between

Russia and Great Britain by an insane act of naval amicide. Identifying a British trawler fleet as a flotilla of Japanese torpedo boats, the Russian gunners let fly with everything they had. The trawlers, known as the Gamecock Fleet, were just tiny one hundred-ton vessels out of Hull, carrying a crew of eight and fishing on the Dogger Bank. How the Russians could have identified them as Japanese warships is beyond comprehension. Some Russian ships claimed they had actually been hit by torpedoes, though in the light of dawn, there were no signs of any damage. It was another case of mass hysteria. Some sailors on the battleship *Borodino* actually donned lifebelts and jumped overboard; others lay prone on the decks with their hands over their ears. Some even ran around the decks wielding cutlasses and shouting that the Japanese were boarding them. One officer bellowed: "Torpedo boats! A torpedo attack! The destroyers! We're done for!" Meanwhile, the big guns of the battleships kept firing at the trawlers, damaging four and sinking one, as well as inflicting hits on each other—the cruiser *Aurora* had four hits below the waterline, and its ship's chaplain was cut in half by a shell. As soon as Rozhestvensky realized his mistake, he was like a man possessed, trying to set things right by throwing one of his own gunners overboard because he continued to fire at a damaged trawler. Aboard the British trawlers, there was only a sense of unreality. One minute they had been peacefully fishing, the next moment they were in the midst of a battle with the Russian Baltic Fleet. Some British sailors tried to demonstrate their peaceful intentions by holding up fish to the Russian searchlights. The boatswain of the trawler *Swift* reported: "Me and the rest of the crew held up fish to show who we were. I held up a big plaice. My mate, Jim Tozer, deck hand, showed a haddock." But such good sense had no effect on the Russians. Two seaman on the *Crane* were decapitated by Russian shells, while the boatswain had his hands blown off.

The truth only dawned with the coming light. It had been a night of madness, with seven battleships in line firing at the cruisers *Aurora* and *Donskoy*. Fortunately, the gunnery had been deplorable—the *Oryol* fired over five hundred shells without scoring a hit. But how would the British react to this unprovoked assault on their fishermen? As the trawler fleet returned to Hull, the whole city turned out to welcome them and mourn their losses, and flags were flown at half-mast. The British press worked itself into a fury of jingoistic hatred for Russia and demanded war. *The Times* wrote:

> It is almost inconceivable that any men calling themselves seamen, however frightened they might be, could spend twenty minutes bombarding a fleet of fishing boats without discovering the nature of their target. It is still harder to suppose that officers wearing the uniform of any civilized power could suspect they had been butchering poor fishermen with the guns of a great fleet and then steam away without endeavouring to rescue the victims of their unpardonable mistake.[51]

Many British officials believed that there was evidence to suggest that some of the Russian officers had been drunk at the time of the attack. One reporter, the famous novelist Edgar Wallace, managed to interview a steward aboard one of the Russian battleships, who said that he had been in the ship's pantry at the time of the attack and had heard a midshipman run into the officer's mess, shouting: "The Japanese are attacking us!" The officers apparently rushed up on deck, though minutes later they sent a request for their glasses and bottles of brandy to be brought up as well. The steward found them there in every kind of disorder, waving swords, firing small arms, and lying full-length on the decks, arguing.

Although the Russian government made an official apology, Britain was slow to forgive. Twenty-eight British battleships of the Channel Fleet raised steam and prepared for action, while swarms of British cruisers shadowed Rozhestvensky's fleet as it moved fearfully across the Bay of Biscay and down the coast of Portugal. The diplomatic storm caught up with Rozhestvensky at Vigo, where he was ordered to leave behind those of his officers who had been responsible for the attack on the trawlers. Some Russian officers regretted the fact that war had not broken out with Britain on the curious grounds that they might as well have been sunk by the British in the Channel as have to go eighteen thousand miles to be sunk by the Japanese.

The extraordinary instances of naval amicide that occurred during the voyage of the Russian Baltic Fleet almost defy analysis. The usual factors that apply to friendly fire incidents are almost all present, but even these hardly account for the madness of the Dogger Bank incident. After early Japanese successes in the war, the Russians appear to have attributed to the Japanese sailors almost superhuman qualities and certainly far greater efficiency than they themselves possessed. Apart from Admiral Rozhestvensky, neither officers nor men had any confidence in their ability to carry out their task, and the entire fleet seemed doom-laden from the outset. In a sense, they were right. In terms of both ships and men, the Japanese were superior, yet there can be few instances in all military history where one army—or navy—went into action so certain of defeat. As a result, fear caused by an overestimation of the enemy resulted in a kind of collective inferiority complex. This sapped morale, until minor incidents were magnified in importance by the Russian crews and used to confirm their own low opinion of themselves. This low morale combined with a natural fear of the enemy and a belief that he was able to threaten their ships even eighteen thousand miles away from his home bases, made the Russian sailors look for him in the most unlikely places, like the North Sea or the Baltic. Few of Rozhestvensky's men seem to have considered the impossibility of Japanese torpedo boats operating so far from home. And the tsar's use of spies rebounded on his own head, as men like Captain Hartling just fed the fears of the Russian crews with unlikely if not downright ridiculous stories of Japanese cunning.

The threat from U-boats during the First World War came close to defeating Great Britain. Various methods were tried to combat the submarine menace, one of the most original being the introduction of Q-ships—merchantmen with concealed guns—which had a big initial impact. But after a period of success, U-boat skippers became increasingly suspicious of merchant ships that seemed too confident, so that by the end of the war, their successes were few and difficult to achieve. The Q-ship *Cymric* had been a schooner prior to being converted, and on October 15, 1918, it was cruising off the coast of Yorkshire. It had already had one piece of excitement that day when it had almost fired on one of the famous K boats as it lay on the surface, oblivious to its surroundings. Disappointed not to have a U-boat under its guns, *Cymric* sailed on, doubly determined to find a target. Suddenly, a second submarine was sighted. The *Cymric*'s crew went to action stations, but again there was the frustration that it was "one of ours."

Meanwhile, the British submarine J6, commanded by Lieutenant Commander Geoffrey Warburton, was on patrol in the same area. Because the J class submarines slightly resembled U-boats, the crews always took great pains to keep the huge "J6" that was painted on the conning tower bright and clean. But this time, luck was against them. By a curious chance, something was hanging from the conning tower that provided a second upright on the "J" turning it into a "U." Thus, to all intents and purposes, "U–6" was patrolling on the surface when it encountered *Cymric*.

J6 slowly approached the schooner, but unknown to the submarine crew, the Q-ship's gunners were waiting their chance. Keen eyes aboard the *Cymric* noted a limp ensign flying from the submarine's flagpole, but they took no notice of that—they had been fooled by false colors in the past.

Suddenly, *Cymric* broke out the white ensign, dropped its disguise, and opened fire on J6 from point-blank range. The submarine never had a chance. An officer on J6 raised a rifle to fire an identification signal, but it was blown from his hand and the signal was never sent. The Q-ship's four-inch gun scored an immediate hit, while on the submarine an officer was waving something that resembled a large white tablecloth. Before *Cymric* could fire again, the submarine drifted into a fogbank. But as it was swallowed up in the gloom J6's signaler managed to flash out the word "Help."

The Q-ship was still not sure what to make of the mystery submarine and followed it into the fog. There it found J6 sinking, with men jumping over the side and being collected in a small boat. As some survivors reached *Cymric*, their hatbands displayed the words "HM Submarines," and the truth at last dawned on the Q-ship's crew. It was a sickening moment for all concerned. Only fifteen of the submarine's crew survived their ordeal. The subsequent court of inquiry absolved the Q-ship from blame and recorded the incident as a "hazard of war"—a case of mistaken identity and of naval amicide.

• • •

So impressed were the British admiralty with the success of German U-boats in the early months of the First World War that they decided to respond by building their own new submarines to a revolutionary design, making them the largest, fastest, and most powerful underwater craft in the world. The new ships would be revolutionary in that they were to be driven by steam engines and would use two funnels. Unfortunately, once built, these new super-submarines were to enjoy a record of failure second to none. Unofficially they were known as the "suicide club," and they were involved in more damage to themselves and other British ships than to the enemy. On the only occasion that one of them actually fired at a German ship, it scored a direct hit but its torpedo failed to explode.

The new K class was the brainchild of the director of naval construction, Sir Eustace Tennyson-d'Eyncourt. It was clear to men like Winston Churchill and Admiral Sir John Fisher that by 1914, Britain was far behind the Germans in submarine construction, and that something special was needed to redress the balance. The K ships were certainly special, designed to be three times the size of existing British E class submarines, and fast enough to accompany the Dreadnought battleships on their sweeps in the North Sea. Fisher, the first sea lord, demanded a submarine capable of at least twenty knots on the surface, and Tennyson-d'Eyncourt told him that to achieve this, the submarine would need to be driven by steam turbines and carry funnels that could be sealed off when the ship dived.

Not everyone agreed with Fisher that submarines could work effectively on fleet maneuvers. Commodore Roger Keyes, for one, regarded the idea as absurd. In the opinion of Keyes, the Battle of the Heligoland Bight had proved that the British submarines involved were a serious liability. Three times, British submarines mistook British ships for German and tried to torpedo them—a torpedo from one only just missed sinking the cruiser *Lowestoft.* Conversely, three British cruisers tried to ram British submarines, mistaking them for U-boats. In Keyes's own words: "The submarines had proved that they could not be trusted to work in cooperation with surface craft and take care of themselves." But Fisher refused to be deterred. He continued his search for the fast submarine, and by May 1915, he was certain that he had found it. Vickers was ordered to begin work on the K class, based on d'Eyncourt's original design. Although Fisher was shortly to leave the admiralty, a total of fourteen of these steam-driven submarines were ordered before he left.

The K-class submarines were undoubtedly impressive: longer than a football field and heavier—at twenty-six hundred tons—than the latest destroyers. They carried two four-inch guns and one three-inch gun, two five-foot high funnels, twin thirty-foot periscopes, and two tall, retractable wireless aerials. In fact, they carried so much on top that it seemed a miracle that they could dive at all, or resurface. To the professional submariners, there seemed to be "too many damned holes" in the enormous structure, with hatches and vents and funnels, all of which needed to be watertight when submerged. The smallest weakness in any one of the seals would spell disaster.

But the admiralty was delighted. Even before the early models had undergone trials, seven more were ordered during 1916 at a total cost of £6 million. K3 was the first of the submarines to begin its trials. Apart from the fact that its boiler room nearly boiled the men working in it, the heavy sea cracked the windows of the wheelhouse, and a British patrol boat opened fire on it, all went quite well. Unfortunately, when K3 tried to dive—with the future King George VI aboard—it went out of control and sank to the bottom of Stokes Bay, near Portsmouth, with its bows stuck in the mud and its stern rearing high above the water. Fortunately, no lives were lost, and it was subsequently refloated. In January 1917, K3 was sent north to join the Grand Fleet at Scapa Flow. In a heavy swell in the North Sea, water entered both funnels, extinguishing the boiler fires and filling the boiler room to a depth of four or five feet. Only the presence of a subsidiary diesel engine—which Fisher had insisted should be fitted—got the submarine started again and prevented it from sinking.

The tragic case of the ominously numbered K13 was a clear indication of vital design faults. On January 29, 1917, Lieutenant Commander Godfrey Herbert, commanding K13, gave the order to dive in a Scottish loch. The boiler room flooded and the submarine settled on the bottom, just fifty feet below the surface. After a prolonged rescue operation, Herbert was himself saved, but twenty-five of his crew lost their lives. It should have provided a salutary lesson.

For much of 1917, however, the admiralty practiced the "habits of self-deception" on the question of the K boats. Reputations were at stake as well as big contracts, and so the truth was suppressed. Between January and May 1917, the first thirteen K boats all underwent trials, and every one of them experienced trouble. K2, for example, suffered an internal explosion and fire. With no fire extinguishers aboard, the submarine had to surface and extinguish the fire with buckets of seawater passed hand to hand. K6 sank on its trials and spent an uneasy time on the bottom before its compressed-air system was repaired and it was refloated; K4 ran aground; K14 developed leaky plates and electrical fires; yet in spite of these failings, all were sent north to join the Grand Fleet at Scapa Flow. Even the simplest able seaman could have told their lordships of the admiralty that in action a submarine needed to be able to dive in thirty seconds, not five minutes like the K boats. And the fifteen to twenty minutes it took to work up a head of steam meant that the K boats would be sitting ducks for German destroyers.

In June 1917, the Twelfth Submarine Flotilla—K1, K2, K4, K7, and K8—took part in an antisubmarine sweep in the North Sea known as Operation BB, accompanied by destroyers and conventional submarines. They were soon in trouble. In ten days of operation, the British destroyed no U-boats while the Germans sank no less than nine British merchantmen under the eyes of the fleet. K7 was once identified as a U-boat and chased and depth-charged by two British destroyers. Escaping by the skin of its teeth, K7 itself spotted a U-boat and fired a torpedo at point-blank range that hit the German sub amidships but failed to explode. The German submarine promptly dived—something K7 could not do—and escaped. K2, meanwhile, had been reported lost with all hands. The Fair Isle lighthouse claimed to have seen K2 strike a mine and sink. The admiralty responded by sending out telegrams to the next of kin. Two days later, in darkness, an unidentified submarine entered Scapa Flow, setting off a general fleet panic. It was K2. There had been no mine; the lighthouse keepers had merely seen K2 fire one of its four-inch guns before diving. During the same operation, K1 ran aground, its captain escaping censure on the grounds that rats had eaten part of his sea chart. But there was no happy ending for K1. Later in 1917, it was accidentally rammed by K4 and so badly damaged that it had to be scuttled.

Rear Admiral Ernest Leir described his experiences with the K-class submarines in this way: "The only good thing about K boats was that they never engaged the enemy." Yet the K boats did undergo one trial by combat—an amazing instance of naval amicide, ever afterward known as the "battle" of May Island. In December 1917, the K boats were moved from Scapa Flow to Rosyth, and Admiral Beatty decided to use them in an important operation that would include battleships, battle cruisers, light cruisers, and the two flotillas of type K submarines. On February 1, 1918, Operation EC1 took place. Commander Leir—as he then was—in the cruiser *Ithuriel* led the five K boats of the Thirteenth Flotilla along the Firth of Forth, directly in the wake of the battlecruiser *Courageous,* flying the flag of Vice Admiral Sir Hugh Evan-Thomas. Some five miles behind him came the four battle cruisers of the Second Battlecruiser Squadron, and behind them the cruiser *Fearless* leading the four K boats of the Twelfth Flotilla. The night was clear, the sea was calm, and there seemed little reason to expect trouble.

But nemesis approached in the shape of eight armed trawlers sweeping for mines in the firth. They operated out of May Island, and through a breakdown in communication, neither they nor the officers involved in Operation EC1 were aware of the others' operation. A mist now descended that reduced visibility so that the *Ithuriel* lost contact with the battle cruiser *Courageous* ahead. As *Ithuriel* began to lose its way, chaos ensued among the five K boats. Minesweeping trawlers appeared out of the mist, flashing their navigation lights and baffling the K boat commanders who were trying to follow the stern light of the *Ithuriel.* K14 tried to go hard a'starboard, but its helm jammed for six minutes and its commander had to stop engines to avoid going around in circles. Suddenly, at nineteen knots, K22 came crashing into the virtually stationary K14. Lights now flashed out from all directions—the signalman on

K14 using an Aldis lamp to call for help. Meanwhile, the four huge battle cruisers were bearing down on the scene of the collision, oblivious to the situation ahead. The *Australia* passed by the collision safely, managing to detach a destroyer to investigate, but the last of the big ships—the *Inflexible* — plowed straight into K22. Ironically, K22 was the renamed K13 that had sunk so tragically in the loch, with the loss of many lives.

Meanwhile, Leir in the *Ithuriel*, with his remaining K boats, was responding to the appeals for help from K14 and K22. He was now virtually at right angles to the approaching battle cruisers. Although he was able to maneuver the cruiser out of harm's way, the submarines were too sluggish to move quickly in any direction. In the event, *Australia* and its sisters just managed to scrape past K12 with inches to spare.

Now steaming up the firth came the cruiser *Fearless*, bringing the other four K boats at their full speed of twenty-one knots. With a horrible inevitability, they joined the confusion ahead. *Fearless* rammed into K17, sending it straight to the bottom. As *Fearless* reversed its engines, *Ithuriel* and K11 rushed back to the scene of chaos to look for survivors from K14 and K22. Even now the disasters of that early morning were not over. Traveling at full speed, K6 rammed into K4, which, for some reason, was unlit and stationary across its path. K4 was cut in half and sank rapidly, and K6 narrowly avoided being dragged down with it.

The K boats now faced one more adversary—and the most dangerous yet. At the tail of the British line were the huge battleships of the Fifth Battle Squadron, with their escorting destroyers. While frenzied efforts were made to rescue the survivors of the damaged and sinking K boats, the destroyers escorting the battleships arrived and cut straight through the scene, washing away or cutting to pieces the survivors from K17. Over a hundred men lost their lives that morning.

Roger Keyes had been right in 1914. The verdict of the court of inquiry that followed the disaster was that submarines could not safely work with surface craft. There had been incompetence by individuals, but on such a scale that it was impossible ever to get to the bottom of it. The main fault lay with the K boats themselves. As their crew said: "they were killers." Strangely, even when the war ended, interest in the K boats did not wane. More were built—only to fail. On January 20, 1921, K5 sank with all hands while on exercises. Six months later, K15 sank in Portsmouth Harbor. K22—raised from the Scottish Loch in 1916 and damaged at May Island in 1918—survived to achieve the feat that everybody had expected eventually. Off the west coast of Scotland in 1921, it dived with both funnels still open.

The problem of friendly fire that had been a frequent factor in operations during the First World War had not been overcome by the time war began in September 1939. Within a week of the hostilities breaking out, the Royal Navy had lost its first submarine, and as far as anyone could tell, no German vessel had been involved. Tragically, it was the continuing problem of friendly fire. The telegrams sent by the admiralty to the next of kin, interestingly enough, still contained the soothing untruth that their loved ones had been "lost in action." It was just a way of shielding the bereaved from the feeling of utter waste that went with the knowledge that one's menfolk had been killed by their own colleagues.

On September 9, 1939, five British submarines, including the *Oxley*, commanded by Lieutenant Commander Bowerman, and the *Triton*, commanded by Lieutenant Commander Steel, took up station off the coast of Norway. The submarines were stationed twelve miles apart, and each had a sector to patrol, *Oxley* sector 4 and *Triton* sector 5. The two submarines had established good communications earlier in the day. Unfortunately, *Oxley* had begun to lose its bearings and unknowingly stray into *Triton*'s sector. Just before midnight on September 10, one of *Triton*'s officers sighted an unidentified submarine through his binoculars. He called Steel to the bridge, and the crew went to action stations. Steel admits that the thought of it being *Oxley* passed through his mind, but he dismissed it, as he had been in touch with it that afternoon and had given it *Triton*'s position. If it was *Oxley*, it was six miles off its station, which was surely impossible.

Steel now took the precaution of locking his armament on the intruder and then signaled a challenge to it. There was no reply, even after another challenge was made. Steel now ordered a signal grenade to be fired, and soon three green lights illuminated the scene. Still there was no reply. Steel had done what he could; the submarine must be hostile. At once, he ordered two torpedoes to be fired and was gratified to hear an explosion, notifying success. The enemy submarine disappeared, leaving just two survivors swimming toward the *Triton*. One can barely imagine the shock the *Triton* crew experienced when they pulled out of the water Lieutenant Commander Bowerman of the *Oxley*, along with just one able seaman. The rest of *Oxley*'s crew were lost.

Nobody, either at the time or since, blamed Lieutenant Commander Steel for firing. What baffled everyone was the *Oxley*'s failure to respond to any of Steel's signals or requests for identification. The answer was a mixture of misfortune and incompetence. Apparently, Bowerman had thought himself two miles inside his own sector at a time when he was in fact four miles inside *Triton*'s. He had been called to the bridge when *Triton* fired its signal grenade, but when he tried to reply, his grenade malfunctioned. His bridge officer, Lieutenant Manley, claimed to have answered *Triton*'s challenge, but Bowerman was not convinced that this had been done properly. Before he could put things right, his ship was struck by *Triton*'s torpedo and he was catapulted into the sea. There was little more to be said. *Triton* had acted correctly throughout, and *Oxley* had been lost through bad luck and slack work by its bridge officer. Nobody could have allowed for the malfunctioning of the signal grenade, but if Manley had replied initially to *Triton*'s challenge, the tragedy need never have happened. In wartime, nobody gives you a second chance; *Triton* had shot to kill.

On February 22, 1940, an amazing blue-on-blue occurred in the North Sea, and the culprit was a lone German bomber. Its achievement, in other circumstances, would have been memorable, but in sinking two German destroyers at the cost of 578 German lives, it was one unlikely to win the perpetrators any honors.

The German First Destroyer Flotilla of six fast, modern destroyers under the command of Fritz Berger in the *Friedrich Eckoldt* set out on a mission to intercept the British trawler fleet operating on the Dogger Bank. It seemed a simple job on paper, and Navy Group West (which was controlling the operation) received assurances from the Luftwaffe that "air reconnaissance" would be available to "cover the destroyers' departure and return." In addition, apparently, "Bomber forces will also be at readiness." It all sounded very reassuring, even if nobody quite understood the relevance of this last reference to bombers. Nevertheless, on the assumption that it was referring to the bomber support that would be available if the destroyers ran into formidable opposition—an altogether unlikely event—the destroyers sailed.

At 1913 hours, Berger's flagship detected the sound of an aircraft overhead, apparently a bomber. The immediate problem was: Was it friendly or not? The plane made no attempt to identify itself, and so two of the destroyers fired a few random shots, whereupon the plane returned fire with its machine gun. Then, as the moonlight illuminated the side of the plane, a lookout on the *Max Schultz* called out: "It's one of ours," claiming to have seen the German insignia. But was he mistaken? On the *Friedrich Eckoldt,* Berger was certain that if German bombers had been in the vicinity, he would have been told. Suddenly, at 1944 hours, the *Leberecht Maas* radioed that it had spotted the plane again, coming out of the clouds. On his radio receiver, Berger heard a dull explosion, followed by another. The plane had dropped two bombs, hitting the *Leberecht Maas* amidships. Berger immediately turned back to help his struggling colleague.

What was actually happening, unknown to Berger and his men, was that the Tenth Air Corps had decided to carry out an attack on enemy shipping in the North Sea on the same day that the destroyers began their own operation against the trawlers. Navy Group West was accordingly informed that two squadrons of Heinkel He-111 bombers would be hunting British shipping that night, but someone forgot to warn the destroyers. Unfortunately, one of the Heinkels, flown by Warrant Officer Jäger, became detached from the others. Jäger's observer, Sergeant Schräpler, spotted a ship below that he later described as "rectangular" and clearly a merchantman—in fact it was one of Berger's destroyers. When asked at the court of inquiry how much experience he had had in identifying ships from the air at night, Schräpler replied: "None at all. That was the first time." Both plane and destroyers were unaware of each other's identity owing to a simple breakdown in communications between the Luftwaffe and the navy—and the consequences were severe. Convinced that the ships were hostile, Jäger decided to attack and struck the *Leberecht Maas* with two bombs.

Commander Berger, meanwhile, raced back to help, only to find that the *Leberecht Maas* had broken in half and was sinking. He immediately ordered the other destroyers with him to lower their boats and try to rescue the survivors. Mysteriously, a few minutes later, there was another explosion and a bright flash of light in the darkness. Captain Böhmig of the *Theodor Riedel*, engaged in trying to rescue the men swimming in the sea, suddenly heard the message: "Sounds of submarine, strength five decibels, to starboard." So there was a British submarine, after all, Böhmig thought. It must have been that which sank the *Leberecht Maas*. At once, he ordered depth charges to be launched, but his own ship was almost shattered by their explosions, damaging its steering mechanism and making it go around in circles—at the mercy of the submerged British submarine. The sound of depth charges being fired convinced the other German captains that Böhmig must have located a submarine. Meanwhile, lookouts on the *Erich Koellner* called out that they had also spotted a submarine. As Berger later explained: "I was by now convinced that a submarine was present . . . to date British bombers had failed to score a single hit even by day, was it likely that they had suddenly scored a series of hits by night?" So, while Berger and his captains scurried about in the dark looking for a submarine, it was a "red-letter day" for Jäger and his crew, having sunk two enemy vessels in one night.

Berger, fearing that his ships might be sitting ducks during a rescue operation, made the heartbreaking decision to abandon the men in the water and to leave the area at full speed. So quickly did the destroyers leave that one ran over its own rescue launch, killing its own men and drowning the recently rescued survivors from the *Leberecht Maas*. Signals now reached Berger thick and fast. In the darkness, his crew panicked, sighting submarines all around and torpedo tracks making toward them as if drawn by a magnet. What they could see, in fact, was their own bow wash illuminated by the moon. Hysterical lookouts next claimed to see a conning tower that one of the destroyers tried to ram, only to find at the last moment that it was the bow section of the doomed *Leberecht Maas,* slipping beneath the waves.

Berger was by now thoroughly confused. He had just come across what he believed was the stricken *Leberecht Maas,* even though he had been heading in the opposite direction, away from it. And then the truth struck him: A second destroyer had been hit and was sinking—the *Max Schultz.* There must be a submarine—perhaps more than one. It was his task to save the remaining four members of his squadron. The survivors from his two wrecked ships would have to be abandoned. With a heavy heart, he called off the rescue operation and ordered the destroyers to head for home. Nobody from the *Max Schultz* survived, and just sixty from the *Leberecht Maas* were pulled from the sea.

The disaster that struck Berger's flotilla was a combination of friendly fire and sheer hysteria. There had been just one bomber—Jäger's HE-111—which had achieved the remarkable tally of sinking two destroyers with just five bombs—and at night as well. There had been no British submarines at all. All sightings were the products of minds confused by their own bow waves and the effects of moonlight through mist.

Jäger, meanwhile, had every reason to be satisfied with his haul. He reported his first success as a "3,000-ton steamship." Unfortunately, the position he gave for his kill was exactly the same as that currently occupied by Berger's flotilla. Rear Admiral Otto Ciliax was immediately concerned, yet how could the pilot confuse a single merchant ship with six destroyers? One can only guess why Jäger's observer made such an elementary error. In his excitement, he showed as little control as his naval colleagues below, who saw periscopes, conning towers, and torpedo tracks where none existed. At last, the truth began to materialize. However unpalatable the news was, Hitler had to be informed, and he demanded an immediate inquiry. Enemy action was conclusively ruled out. It had been an administrative error, a tragic and unnecessary example of friendly fire and of failures in coordination between members of different services.

If the German Luftwaffe had blundered, the British RAF was

not slow in following them. During the hunt for the German battleship *Bismarck* in May 1941, a serious attack was made by Swordfish torpedo bombers from the British aircraft carrier *Ark Royal* on the British cruiser *Sheffield,* which was incorrectly identified as the German battleship. Radar sightings of a large warship on approximately the right setting convinced the fifteen bombers that they had the *Bismarck,* fresh from its victory over the *Hood,* in their sights. Determined to avenge the loss of Britain's greatest warship, the Swordfish pilots were perhaps more eager to strike than to identify their target. Fortunately for the *Sheffield* and its crew, faulty fusing caused the torpedoes of the first wave of Swordfish to explode on contact with the sea. But the second wave fired their torpedoes, which ran true, and for several tense minutes, the cruiser had to maneuver wildly to avoid them. A third wave of Swordfish did not release their torpedoes, as a positive identification of the cruiser had been made by that time. It was an object lesson for naval air pilots in the importance of accurate identification of targets when there are a large number of friendly vessels in the area.

In 1943, misidentification led to the sinking of an American transport by a friendly PT boat, whose officer had relied on assurances that there were no friendly vessels in the area instead of believing the evidence of his own eyes. And when an American submarine torpedoed the Japanese merchant vessel *Arisan Maru* in October 1944, it scored an appalling blue-on-blue, as the ship was conveying thousands of Allied prisoners of war from the Philippines to Japan. Everyone was drowned.

Weapons systems are generally designed to operate within a range of climatic conditions. However, when they are exposed to conditions far beyond the norm, there is always the danger that they will malfunction. If the weapon is offensive in character rather than defensive, then the worst that can usually happen is a failure to strike the enemy. However, there is one recorded occasion of a torpedo turning a full 180 degrees and inflicting decisive damage on the ship that fired it. The victim in that instance was the British cruiser *Trinidad.*

On March 29, 1942, *Trinidad* was escorting convoy PQ13 in Arctic waters when it was attacked by three German destroyers.

It was so cold that spray froze instantly as it landed on the decks. After an exchange of gunfire, *Trinidad* fired three torpedoes at the Z26, but two of them were so iced up that they failed to leave their tubes. The third torpedo malfunctioned when the oil in its motor and gyroscope froze, causing it to change direction, swing around, and return the way it had come. The torpedo hit the cruiser amidships, damaging it severely. It was only with extreme difficulty that *Trinidad* limped into Murmansk for repairs, victim of a self-inflicted wound.

5

Discipline and Friendly Fire

Fear is the natural state of the soldier. We know from ancient writers that Greek warriors often urinated involuntarily as the enemy battle line bore down on them. Their natural inclination was to run away, to avert the danger that was before them, but most, through an exercise of willpower, stayed in line and fought. It was not that these men who did not run had not wanted to, but that they had been able to control the instinct for self-preservation. Some factor, perhaps a greater fear of failure or disgrace, or of letting their comrades down, had kept them in the place of danger. They had shown courage, but as we know from modern research, courage is no absolute. On another day, in another place, a "brave" man can run away. In his book *The Anatomy of Courage,* Lord Moran has written: "Courage is willpower, whereof no man has an unlimited stock; and when in war it is used up, he is finished. A man's courage is his capital and he is always spending. The call on the bank may be only the daily drain of the front line or it may be a sudden draft which threatens to close the account." In 1914, it was the task of military discipline to support a soldier's own self-discipline, so that when the latter began to waver, he found support in the chilling imperative of the gun pointed at his back or at his head. From that point onward, war consisted for each individual soldier of balancing a chance against a certainty: the chance of death or mutilation at the hands of the enemy against the certainty of death and disgrace at the hands of his own officers; a grisly form of friendly fire that has rarely earned more than a footnote in military memoirs.

In his important book *Morale*, John Baynes has shown that military discipline has always had two purposes. In the first place, as we have seen, it prevents a soldier giving way in circumstances of great danger to a natural instinct for self-preservation. It shows him where his duty lies, even if doing that duty should cost him his life. In the second case, it maintains order within the army itself, so that neither in its parts nor as a whole does it abuse its power. It is the first aim that most concerns the writer on friendly fire, for it is in situations where the instinct for self-preservation has prevailed over the soldier's duty that it is necessary for military discipline to reveal itself in all its stark simplicity. As one writer has put it: " 'The avenue to the rear' must be absolutely closed in the soldier's mind, and military discipline serves to buttress the individual in his struggle with his own fears."

An American private from the time of the Civil War spoke for every soldier who has ever gone into battle, when he wrote: "The truth is, when bullets are whacking against tree trunks and solid shot are cracking skulls like egg shells, the consuming passion in the heart of the average man is to get out of the way. Between the physical fear of going forward, and the moral fear of turning back, there is a predicament of exceptional awkwardness, from which a hidden hole in the ground would be a wonderfully welcome outlet."

Every soldier needs to overcome such fear in battle if he is to be of any military value at all. Successful armies throughout history have been those that have overcome this fundamental problem. The Romans, for example, exercised a severe military discipline, as Gibbon described:

> . . . it was impossible for cowardice or disobedience to escape the severest punishment. The centurions were authorized to chastise with blows, the generals had a right to punish with death; and it was an inflexible maxim of Roman discipline, that a good soldier should dread his officers far more than the enemy . . . In his camp the general exercised an absolute power of life and death; his jurisdiction was not confined by any forms

of trial or rules of proceeding, and the execution of the sentence was immediate and without appeal.[52]

In the army of the Prussian king Frederick the Great, the Roman ideal was imitated, perhaps even exceeded. Frederick himself followed Gibbon by saying that the common soldier must fear his officer more than the enemy. Prussian officers would strike their men for even the most trivial lapses, inflict tortuous field punishments like riding the wooden horse, brand or flog them, or make them run the gauntlet. For more serious offenses, there were capital punishments, such as shooting, hanging or breaking on the wheel. Barbaric as this may sound, it was not a regime confined only to the Prussian army. The British redcoats of the eighteenth century were known throughout Europe as the "bloodybacks" as a result of the British predilection for flogging. As sophisticated an officer as James Wolfe ordered that "a soldier who quits his rank, or offers to flag, is instantly to be put to death by the officer who commands that platoon, or by the officer or sergeant in rear of that platoon, a soldier does not deserve to live who won't fight for his king and country." A century and a half later, in 1914, these identical words could have been spoken by many British officers on the Western Front, charged with transforming a massive civilian army into an efficient military machine. As one of them wrote: "I, who am a soldier, know that it is difficult to leave the shelter of a shell-hole for a final rush in the face of a deadly shower of bullets and the certain knowledge that cold steel awaits. It is less difficult, however, if there is the knowledge that a loaded revolver for use against the enemy is also loaded for use against you if you fail to jump forward when the barrage lifts."

John Baynes cites an example of a British soldier panicking in 1915, shouting to the men around him, "Get out! Get out! We're all going to be killed" just as the Germans were launching a counterattack. A sergeant promptly split his skull with a shovel he was holding, and the panic ended instantly.

In 1930, a senior British officer, Brigadier General Frank Percy Crozier, caused an outcry in Britain by telling what he claimed to be the "unvarnished truth" of the First World War. In two books, the second provocatively entitled *The Men I*

Killed, Crozier spoke of the way in which the British troops had been kept at their posts, not, as the popular press had always told the British people, because they liked a "good scrap" or were eager to punish the Huns for their atrocities in Belgium, but because they feared that if they did not go forward to face the German guns, they would be killed by their own officers. Crozier elaborated on this theme by writing openly of the officers and men he had personally shot. It was at his command, Crozier claimed, that fleeing Portuguese troops were machine-gunned by their British allies in 1918. In these books, it was Crozier's intention to shock and to force a complacent public to face the true cost of war and of victory. Nevertheless, by opening up the subject for general discussion, Crozier was also raising the whole question of discipline in wartime. Was it acceptable, in the pursuit of victory or the avoidance of defeat, for men to shoot their colleagues in the heat of battle or for the authorities to execute soldiers for cowardice and desertion? This aspect of friendly fire is every bit as worthy of investigation as incidents of an accidental nature.

Brigadier General Crozier's claim that British officers were sometimes shot for cowardice in the thick of battle was a startling revelation at a time when courage was not just an attribute of military rank but a *sine qua non* for a whole social class in Britain, that had ruled unchallenged for centuries. When Crozier wrote his books, the truth about executions of British soldiers had not reached the general public, so that the idea of discipline imposed at the point of a gun was particularly shocking. It may have been believed of the German or Russian armies, but not of the British. Yet Cozier made no apologies for what he had done:

> Strictly from the military point of view I have no regrets for having killed a subaltern of British infantry on that same morning I ordered our machine-guns and rifles to be turned on the fleeing Portuguese. It happened on the Stratzeel Road. It was a desperate emergency. I had to shoot him myself, along with a German who was running after him. My action did stem the tide; and that is what we were there for.
>
> Vividly I still remember that scene. It might almost have been only yesterday. Never can I forget the agonized expression on

that British youngster's face as he ran in terror, escaping from the ferocious Hun whose passions were a madness and who saw only red.

... Oh, I know you will ask why I killed that British subaltern. The answer is more obvious than easy. My duty was to hold the line at all costs. To England the cost was very little. To Colonel Blimp in his club and Mrs. Blimp in her boudoir the cost was nothing. To me? Even if the effort did mean murder, the line had to be held.

There were other British soldiers there, the last of a remnant. Panic spreads so easily when the madness of a moment assails you. And a running man is a dangerous madness. Only, you don't stop to think once you've begun running.[53]

Another incident described by General Crozier indicates just how often these battlefield killings took place:

I remember an officer of birth and breeding of the special reserve—a captain, with experience at Gallipoli, it was said—who joined with a reinforcing draft after the Somme battle. By his talk and bearing I imagined him an efficient officer, but being always suspicious of the talking soldier I took the precaution of placing him under test as second-in-command of a company, under a most capable temporary captain of the new armies.

He objected. It was as well. For he turned out to be useless and afraid.

"What did you do exactly at Gallipoli?" I asked him one day, after finding him crawling on his stomach, in daylight well behind the lines, on his way up to his company, in an effort to avoid imaginary danger caused by the cracking of overhead, indirect rifle-fire. The firing had got on his nerves.

"I was on the staff," he replied. And it then transpired that he had got no further than one of the islands and was, in effect, a "bottlewasher." I mentioned this fact to his captain, who told me that he had made up his mind to get the fellow bumped off in a scrap in No-Man's-Land ... It was Montgomery speaking ... who, it was whispered, had found it necessary to slay many of his side in order to restore reason and confidence [to] a reinforcing regiment from Yorkshire.[54]

Official statistics, only recently made available, show to what extent the British military authorities used the death penalty—

or the threat of it—to maintain discipline in their largely civilian armies from 1914 onward. Between August 4, 1914, and March 31, 1920, a total of 3,080 men were sentenced to death under the regulations of the Army Act, of whom 312 were executed. Excepting those men punished for serious crimes like murder, rape, or violence against civilians, the vast majority of those who were shot suffered for purely military "crimes," such as desertion or cowardice. They were punished *pour encourager les autres*, or rather to instill sufficient fear into their fellow soldiers that they would endure the almost unendurable in the knowledge that failure to do so would earn them certain death and ignominy at the hands of their fellows.

At the end of the First World War the German commander, Field Marshal von Ludendorff, spoke with some envy of the severity of British and French military discipline compared with that in the German army: "The Entente no doubt achieved more than we did with their considerably more severe punishments. This historic fact is well established." Certainly the British and French had many more executions than the Germans—the French during their 1917 mutinies shot an undisclosed number of mutineers to restore order—and the evidence from diaries and memoirs builds up a picture of numerous "unofficial" executions during the heat of battle. One French general ordered his artillery to bombard his own trenches when his men refused to leave them, and there are many examples of British officers using death or the threat of death to move their men forward. Lieutenant Colonel Lambert Ward recorded one occasion where a brigade of the Third Division had "cracked to a man. You could not send them back to base, yet they were in such a state that they would willingly have taken ten years' penal servitude to stay out of the line. In these circumstances it was only fear of death that kept them at their posts." Another British colonel was quite open about the fact that he ordered his machine-gunners to fire at some British troops who were in the process of surrendering in April 1918. As he said: "Such an action as this will in a short time spread like dry rot through an army and is one of those dire duties which calls for immediate and prompt action."

A large majority of the British soldiers executed during the First World War died to maintain discipline. Their "crimes" were insignificant in the context of a civilian criminal code—desertion, cowardice, disobedience, striking a superior officer, sleeping at one's post—yet in the context of maintaining morale in an army, the victims not only had to die but had to be seen to die by their comrades. Their punishment was in no way retributive, only exemplary. British military authorities attempted to extend their draconian measures to the all-volunteer Australian force in France after 1916, but the Australian government refused to allow it. Most British officers regarded the Australians as undisciplined, untidy, and disorderly soldiers. On the other hand, they were the best shock troops in the British army, highly rated by their German opponents. Their high morale and low desertion rates were not maintained by fear of the gun or the firing squad, but by a healthy *esprit de corps.* It is significant that no British commander of the period seemed to grasp this fact and seek to re-create it in his own troops.

Early executions of British soldiers in 1914 were all of regulars, who had already seen service in the army before the war. The second man shot, Private George Ward of the First Battalion of the Royal Berkshire Regiment, was executed on the recommendation of his corps commander, Sir Douglas Haig, to act as an example to others. Ward's execution was, in fact, botched. As he was being taken out to be shot, he broke away from the guard and tried to escape, being then shot in the back as he ran. He was fetched back on a stretcher and shot in the head by the sergeant of the guard to "finish him off."

One of the earliest volunteers to be shot was Private John Byers of the First Royal Scots Fusiliers, who had falsified his age to get into the army and was just sixteen when he died. Byers was posted to France just two weeks after enlisting, went absent without leave, and was sentenced to be shot on February 6, 1915. Significantly, Byers's comrades who made up the firing squad were reluctant to shoot him, and it is rumored that he was killed only after a third volley. The effect on the men of the fire party was severe. An officer who had been an eyewitness at the execution later told Labour MP Ernest Thurtle: "When they came back, tough characters though they were supposed to be, they were sick, they screamed in their sleep, they vomited immediately after eating, All they could say was: 'The sight was horrible, made more so by the fact that we had shot one of our own men.' "

The most controversial part of the disciplinary process concerned those men who were suffering from extended exposure to artillery fire, known today as shell shock. As the condition was at that time unknown to medical science, those sufferers who were examined before court-martial were regarded as shirkers rather than men suffering from nervous exhaustion. Rifleman Bellamy of the King's Royal Rifle Corps had been serving in the trenches near La Bassee when the Germans exploded a mine under part of his battalion's trenches. The shock was tremendous, and immediately afterward, Bellamy refused to come out of his dugout and take up his position in the firing line. Bellamy told his sergeant that he felt too shaky to stand on the fire-step. After court-martial on a charge of cowardice, Bellamy was referred to a medical board, but they were unable to find any physical cause for his condition. As a result, on July 16, 1915, Rifleman Bellamy was executed at the village of Le Quesnoy, the first but certainly not the last man on record to fall victim to the psychological effects of shell fire.

The first man to die for striking an officer was Private Fox of the Second Battalion of the Highland Light Infantry. The incident occurred while his unit was in a rest area behind the lines, and revolved around a kit inspection. Fox had only recently received tragic news from his home in Scotland and had been drinking on the night before the offense. On the morning of April 10, 1916, his company commander upbraided him for the dirty condition of his boots, prompting Fox to start swearing and then step forward and kick the officer in the knee. He was arrested on the charge: "When on active service striking his superior officer, being in the execution of his duty." In spite of all the extenuating circumstances, Fox was sentenced to death and shot on May 12. It was a savage punishment that reflected the pressures the army authorities felt in maintaining discipline in a British army growing to an unprecedented size. It is significant that recommendations for mercy by the courts were being ignored by the commander in chief, Sir Douglas Haig, with great regularity at this stage of the war.

In the diary of Captain L. Gameson, a medical officer with the British Fifth Army near Lille in 1918, the officer describes two "accidents" that occurred during his supervision of two separate executions. At the first, of two Cape Coloreds from the First Battalion of the Cape Colored Labor Corps, Gameson had the following close escape:

> The firing party of twelve I think, then took up their positions facing the stakes. They were given final instructions, and words of encouragement which manifestly they needed. The loaded rifles were grounded. The party filed out of sight. A sergeant loaded some of the rifles with blanks—he said—no man would know whether he was partly responsible for the killing . . . As he picked up one rifle it went off. The bullet zipped by my head; it passed, so I estimate, precisely just over my left shoulder where that structure joins the neck. A stray piece of wire, with one end fixed in the ground, had in some way discharged the rifle. "My God, sir!" the scared sergeant shouted, "it nearly got you."[55]

Gameson was even luckier to survive another potentially fatal accident that occurred at the execution of Private F. Alberts on May 13, 1918:

> Again I came unpleasantly near to being shot myself; this time with a revolver.
>
> The squad's aiming had been bad. I dashed in immediately after the volley and at once told the sergeant on the opposite side of the moribund victim that the man still lived. With a revolver, he shot the man through the skull at its thinnest part. I repeat that I stood on the opposite flank of the victim. The sergeant fired before I had any chance to remove myself. The bullet went clean through and buzzed by my head as I bent over the semi-unconscious man tied to his stake. I told the sergeant it was much to his credit that he was so rattled by our sordid work as to be irresponsibly careless; perhaps the firing squad had been rattled.[56]

Since the end of the American Civil War, only one American soldier has been shot for desertion. First World War executions in the United States Army had been for murder, often linked to mutiny, or offenses like rape. Private Eddie Slovik, shot by a firing squad from the 109th Infantry Regiment of the Twenty-eighth Division, died on the morning of January 31, 1945, in the village of St. Marie aux Mines, in Alsace. He was shot because he ran away, like thousands of American soldiers had run away on battlefields in Europe and the Pacific. But he was selected to be an example to others of what would happen to men who gave way to their fears. It was his misfortune to desert at a time when American fortunes were low in Europe, when the battles in the Hurtgen Forest and the Ardennes were stretching the willpower of commanders and there was a need to stiffen morale.

On October 24, 1944 the Twenty-eighth Division had moved in to replace the Ninth Division in the Hurtgen Forest. The Twenty-eighth was commanded by the tough Norman Cota, who was determined to succeed in taking the village of Schmidt in the middle of the forest, which the Ninth had been unable to do. But Cota's tough talking was going to cost the lives of thousands of his men, and the only one whose name would be remembered was the one who funked it—Eddie Slovik. Slovik was going to become famous as the only man shot for desertion by the U.S. Army in World War II.

The Twenty-eighth Division had made a bad start in Hurtgen. The "American Luftwaffe" did what it could to help them by carrying out a bombing of the forest, but all they managed to do was to hit Cota's men, killing and wounding twenty-four of them. Yet when "Dutch" Cota said he meant business, he was not joking, and in just two days, the Twenty-eighth Division took the village of Schmidt. But unknown to Cota, the Germans were planning a counterattack, with Mark IV tanks, immune to their own mines. These tanks powered their way through the forest and ripped into the Americans. Soon everyone was running: It seemed as if the whole division was racing madly through the trees. "It was the saddest sight I have ever seen," said one officer. "Down the road from the east came men from F, G and E Companies: pushing, shoving, throwing away equipment, trying to outrace the artillery and each other, all in a frantic effort to escape. They were all scared and excited. Some were terror stricken." Officers were seen running with the men—even a battalion commander reported himself sick with combat fatigue. It was a catastrophe for the Twenty-eighth Division and for General Cota. And someone had to take the blame. The men had run; one of them must suffer to encourage the others: Eddie Slovik. Eddie Slovik was chosen to take the blame.

On the other side of the hill, the Germans were using their own draconian methods to maintain discipline. During the entire war, they shot ten thousand Eddie Sloviks. Commuting the death penalty for deserters did not provide them with a comfortable stay in prison, out of the fighting. Instead, they were formed into special punishment units and sent on "Ascension Day missions," or suicide runs, from which few ever returned.

During its time in the Hurtgen, the Twenty-eighth Division suffered 45 percent casualties. But even that was not enough for the American high command. As had happened so often during the First World War, soldiers were being sent to fight in impossible terrain because their commanders were too far behind the front and knew nothing of the conditions. At Supreme Headquarters in Paris, it is said Eisenhower was out of touch with events at the front. He appeared complacent, while American lives were uselessly thrown away. Even Eddie Slovik's pitiful appeal to Eisenhower asking for clemency fell on deaf ears. Perhaps he never even read these words:

> *Dear General Eisenhowser* [sic],
>
> *How can I tell you how humbley* [sic] *sorry I am for the sins I've committed. I didn't realize at the time what I was doing or what the word desertion meant. What it is like to be condemned to die. I beg of you deeply and sincerely for the sake of my dear wife and mother back home to have mercy on me. To my knowledge I have a good record since my marriage and as a soldier I'd like to continue to be a good soldier.*
>
> *Anxiously awaiting your reply, which I earnestly pray is favorable. God bless you and in your Work* [sic] *for Victory.*[57]

Instead, Eisenhower steeled himself to do something that had not been found necessary even in the First World War. His confidence had been shaken by the German resistance in the Ardennes. Too many U.S. soldiers thought the war was already over. They needed proof that there was no way home but through victory. So Eisenhower confirmed the death sentence on Slovik. Thousands of Americans had deserted their posts and run away—were running away at that moment—from the enemy, and so Eddie Slovik must stand for all of them.

On January 31, 1945, Private Eddie Slovik was shot. Most of his colleagues present thought he deserved to die. They had no compassion for the men who ran away. As one of the firing squad later commented:

> I think General Eisenhower's plan worked. It helped to stiffen a few backbones. When the report of the execution was read to my company formation, the effect was good. It made a lot of guys think about what it means to be an American. I'm just sorry the general didn't follow through and shoot the rest of the deserters instead of turning them loose on their community. I want my children to be raised to believe that their country is worth dying for . . . and that if they won't fight for it, they don't deserve to live.[58]

FRAGGING

Mutiny in the British or American armies has been a rare event in modern times. Yet mutiny can take many forms, and one that has become more widespread in the twentieth century is known as "fragging," that is, the murder of officers or NCOs who show an overeagerness for seeking contact with the enemy or for undertaking other dangerous missions. During the First World War, young officers—often no more than eighteen-year-old lieutenants, fresh from public school—had to order their men to go "over the top" in the face of a withering enemy fire. It was relatively simple for the officer to be shot in the heat of battle, so that his men could remain in their trenches for lack of anyone to lead them. Statistics of this form of assassination are unavailable for the First World War, yet it remains an open question as to what proportion of the extremely heavy casualties among junior officers were caused by their own men. In the Vietnam War, it has been estimated that as many as 20 percent of officer fatalities were the result of fragging, either through shooting or more often by an apparently misdirected fragmentation grenade.

The murder of unpopular officers, often in action, has been widely documented even as far back as Roman times. Richard Holmes reminds us of the Pannonian mutiny of A.D. 14, during the reign of the Emperor Tiberius. One officer who was killed by his own men was the centurion Lucilius, who was nicknamed "Another-Please" from his unpleasant habit of breaking his vine staff on the backs of his men. Christopher Duffy tells us that during the eighteenth century, wounded Spanish officers were often robbed and murdered by their own men and that even in the well-disciplined English army of the time, an early form of fragging did take place. A famous example from the period involved a major of the British Fifteenth Foot Regiment at the Battle of Blenheim in 1704. The man was an infamous martinet who had treated his men harshly in the past and was thoroughly hated. Aware of this, he addressed his men before the battle, saying that if he should fall in the coming fight, at least let it be by an enemy bullet. A soldier replied that they had more to think about than him at that moment, and so the regiment went into battle. After victory had been won, he turned to his men, raised his hat, and called for a cheer, only to fall with a bullet through his head.

During the Napoleonic Wars, French soldiers were not slow to express their opinions of their officers, and to follow them up if necessary with more direct action. Richard Holmes gives an example of one unpopular Napoleonic general being fired on by his own men, while a general sent to enlist men for the National Guard was actually killed. In 1807, after the abortive British assault on Buenos Aires, Rifleman Harris recounts how many of General Craufurd's men were calling the commander in chief, General Whitelocke, a traitor and looking for an opportunity to shoot him. At the Battle of Quatre Bras, just two days before the Battle of Waterloo, the commander of the Ninety-second Foot, Colonel Cameron of Fassfern, was shot dead by a man he had recently had flogged. In the American Civil War, the democratic and undisciplined nature of many of the troops meant that they were less willing than European soldiers of the time to follow an incompetent officer. When

Colonel Adelbert Ames—a West Pointer of high ability—took over the Twenty-fourth Maine Regiment, his regime of hard drilling so annoyed his troops that even as sound a man as Sergeant Tom Chamberlain wrote to his sister: "I swear they will shoot him the first battle we are in." In fact, Chamberlain was wrong, and the men of the Twenty-fourth grew to admire their colonel when they understood what he was trying to achieve. But many Union officers in particular did not live long enough to win the affection of their men.

During the First World War, one notably unpopular sergeant was dispatched by a grenade down his trousers, as General Crozier relates:

> A British N.C.O. had been bullying some of his subordinates. As there appeared to be no way of dealing with the case there, aggrieved men decided to deal with the matter in their own way. As the essence of crime, from the criminal's point of view, is to leave no trace, they decided to get rid of their tormentor in the manner which they thought most suited to that purpose.
>
> A Mills bomb has a local but very violent explosive effect. They decided that the Mills bomb should therefore be their agent. Pulling out the pin from the bomb which held the lever in check and which, in its turn, ignited the charge which exploded after the lapse of some seconds, one of them—they had previously drawn lots for the job—pushed the bomb down the back of the N.C.O.'s trousers after which they made off at lightning speed to avoid the explosion.
>
> Fortunately the poor man was isolated and entirely alone or others would have been killed—but he, ignorant till too late of what had happened, was, figuratively speaking, hoist by his own petard, for he, of course, became a battle casualty.
>
> There was no trace whatsoever left of this N.C.O. while, of course, there was no evidence available.[59]

In his autobiographical work, *Goodbye to All That,* the poet Robert Graves gives an account of one of the few documented cases of fragging in the British army, which he later retold in his poem, "Sergeant-Major Money." On January 20, 1915, the Second Battalion of the Welch Regiment, to which Graves was attached, was in the trenches near Bethune. The weather was appalling, and to liven their spirits that evening, two soldiers—Lance Corporal William Price and Private Richard Morgan—got drunk and shot the company sergeant-major, a man named Hughie Hayes. In fact, Hayes died by mistake, as Morgan and Price had been intending to kill their platoon sergeant, an unpopular man who had been making their lives a misery.

As Richard Holmes has explained in his book *Firing Line*, fragging is merely a symptom of a general breakdown of discipline and the "phenomenon of collective combat refusal," which, until quite recently, was met with the most draconian of reactions by the military authorities, ranging from the Roman "decimation" of mutinous units to the British use of machine guns on their own troops and those of their allies in 1918. Although fragging is clearly a reaction by the rank and file of an army that has deep roots in military history, it only came to public attention during the Vietnam War. Apparently, fragging reached a peak in 1971, with as many as 333 confirmed incidents and a further 158 possible examples. Even these figures give only a partial idea of the extent of the problem. Richard Gabriel estimates that a more accurate figure would be 1,016 fraggings and suggests that as many as 20 percent of officers killed in the Vietnam War were in fact victims of their own men, evidence, if any were needed, of the most serious large-scale disintegration of discipline in American military history. Yet according to Charles Anderson:

> Every soldier, marine, sailor or airman who fragged a unit leader believed at the time of the incident that he acted with more than ample justification. Such a view may sound incredible now but anyone who has seen combat and perceived what it does to one's thinking can appreciate the extreme difficulty, perhaps even the folly, of making value judgments on the thoughts and actions of men in a combat environment.[60]

In a sense, the problem was compounded by the widespread presence of television cameramen and newspaper reporters in the rear areas, who were willing to take up the grouses that were typical in any army. Soldiers felt that they were not the helpless victims of a military system. Instead, they had access to the whole machinery of civilian life, no more than a phone call away. For the first time in military history, civilian soldiers were able to have direct contact with the public at home, telephoning their folks and in direct contact with their loved ones. This had important consequences for the maintenance of discipline in the American forces. Once a soldier felt that he had a way out through the yearly rotation of troops, he could see light at the end of the tunnel in the way that troops in World War I, serving for the duration, with little hope even of home leave, never could. This had few beneficial effects on military efficiency. Instead, many soldiers failed to identify positively with their units, or their officers, and subsequently lacked the *esprit de corps* that was common in early American wars. It was more a question of counting the days rather than trying to do your best for your company or your friends. This weakened the traditional bonds between soldiers and their small unit commanders. It meant that they were unwilling to take any risks and resented it when an officer was too ready to commit them to dangerous operations.

In turn, this made things very difficult for young, inexperienced officers, who felt themselves dragged in two directions at once. Concern for the welfare of his own men was a fundamental responsibility of any officer, but how did one square this with the responsibility to achieve whatever military task he might be given, even if it was likely to prove costly to his men? This problem was felt most acutely by young officers, whose experience in management was necessarily limited and whose desire to please their superiors was often paramount. This could result in the alienation of the soldiers in an officer's unit, increasing tension on both sides and making fragging a possible outlet for the men's frustration. Certainly, for most American soldiers in Vietnam, an officer's willingness and ability to avoid casualties was a far better guide to his credentials as a leader than any outdated concepts of duty and heroism.

C. W. Bowman of the Twenty-fifth Infantry Division described the arrival of a new C.O. who was bad news from the start:

> Toward the end of my tour, our morale did drop because we got a new C.O. Officers only spent six months in the field before they were moved back to the rear. We got a new captain in, and he was Gung Ho John Wayne. But it wasn't the place to play John Wayne because Charlie was real good at suckering people into ambushes . . . Charlie tried his trick one day and we tried to tell the C.O. what was going on. But he didn't want to hear it, so we were suckered into a minefield. Gary stepped on a Bouncing Betty. It blew his leg off.[61]

Dan Vandenberg expressed his contempt for the fussy officers, who made their men's lives hell over little things. It was officers of this kind as well as the gung ho variety who brought out the worst in the American soldier. Men like this officer sometimes ended with a bullet in the back or a fragmentation grenade for company.

> Some of the very few officers I came into contact with were totally incompetent. They were more concerned with how you looked, how your uniform was, or how your hair was cut. The petty chicken-shit stuff that they tried in the United States they tried to enforce in Nam—which is a way to get a bullet in your back. We had enough to worry about staying alive. As far as shaving, we were lucky enough to have water to drink . . .[62]

Officers who harassed their men, even by issuing malaria pills or insisting that they wear their flak jackets and helmets, were in danger of provoking a violent reaction. Whether this can be construed as amicide is a moot point. That it was deliberate rather than accidental is undeniable, which would tend to separate it from the majority of friendly fire incidents that we have studied in this book. Yet it is also undeniable that the wounding or killing of officers by their own men does qualify as friendly fire in that it was committed by men who were fighting on the same side in the conflict. Moreover, fragging should never be regarded simply as murder. Developing the point that Charles Anderson made above, the men responsible for the fragging are convinced that what they are doing is not a crime but a military necessity. Because of the poor relationship that exists between them and their officer, and because there is no way within the system that they can achieve their aim of removing him without harming him, it becomes necessary to sacrifice the individual in the interests of the unit. Combat conditions provide their own compulsion and their own justification.

AFTERWORD

Victory in war does not always go to the side that makes the fewest mistakes. Neither the Germans nor the Japanese were in the same league as the British and the Americans when it came to blundering during the Second World War. Instead, as John Ellis has shown in his new book *Brute Force*, victory went to the big battalions of industry, not infantry. Amicide may not have been an Anglo-Saxon monopoly between 1939 and 1945, but there are times when one can be forgiven for thinking it was. Fiascoes like the airborne operation over Sicily, Operations Cobra and Totalize, and the bombing of Switzerland are more than simply the inevitable consequences of war on a previously unimaginable scale. From an ordinary GI in the Aleutians firing at shadows in the trees, to the navy flak gunner shooting his fellow Americans as they parachute down over the Sicily beaches, one is dealing with fear and the hysteria to which it so easily can lead. In *Wartime: Understanding and Behaviour in the Second World War*, Paul Fussell claims that American antiaircraft gunners were so shaky that during the early days of the Normandy campaign, their guns had tags attached to their triggers saying: "This gun will only be fired under command of an officer." For Fussell, fear was at the root of a disastrous incident in 1988 when an Iranian airliner was shot down by American navy gunners who feared that it was a hostile combat plane. The result was an amicide of 290 civilians.

Since the Gulf War, there has been a concerted effort in the United States to develop an electronic tagging system to identify friendly vehicles during combat. It is hoped that these gadgets will prevent the sort of blue-on-blue incidents that involved aircraft firing on Coalition forces. It is evidence of a positive attitude on the subject of amicide. However, I contend that the message of two thousand years of friendly fire is that, in the final analysis, it is men who make mistakes, through the stress that war imposes, and that this stress is fundamentally linked with fear on their part: fear of death and mutilation, and fear of failure and humiliation. When men are afraid, they will always shoot first rather than identify a target, or drop bombs too early rather than risk the flak. Mistakes will be reduced when men have less to fear. But then that would not be war, and they would not be men.

Bibliography

The following works have been most useful to me during the writing of this book and I have referred to many of them in the text.

Baynes, J. *Morale: A Study of Men and Courage.* London: Cassell, 1967.

Bean, C.E.W. *Official History of Australia in the War of 1914–18.* Sydney: Angus and Robertson, 1942.

Bergerud, E. *Red Thunder, Tropic Lightning.* Boulder, CO: Westview, 1993.

Bradley, O. *A General's Life.* New York: Simon and Schuster, 1983.

Carlton, C. *Going to the Wars.* London: Routledge, 1992.

Crozier, F.P. *A Brass Hat in No Man's Land.* London: Michael Joseph, 1930.

——. *The Men I Killed.* London: Michael Joseph, 1937.

D'Este, C. *Decision in Normandy.* London: Collins, 1984.

de la Billiere, Sir P. *Storm Command.* London: Harper Collins, 1993.

de Monluc, B. *Military Memoirs.* (ed. I. Roy), London: Longman, 1971.

Duffy, C. *The Military Experience in the Age of Reason.* London: Routledge, 1987.

Eichelberger, R.L. *Our Jungle Road to Tokyo.* New York: Viking, 1950.

Ellis, J. *The Sharp End of War.* London: David and Charles, 1980.

Froissart, *The Chronicles of England, France and Spain.* London: Dent, 1906.

Fussell, P. *Wartime, Understanding and Behaviour in the Second World War.* Oxford: Oxford University Press, 1989.

Graves, R. *Goodbye to All That.* London: Penguin, 1960.

Griffith, P. *Rally Once Again.* London: Crowood, 1987.

Hanson, V.D. *The Western Way of War.* Oxford: Oxford University Press, 1990.

Hastings, M. *Overlord.* London: Michael Joseph, 1984.

Hibbert, C. (ed.), *The Recollections of Rifleman Harris.* London: Leo Cooper, 1970.

Holmes, R. *Firing Line.* London: Penguin, 1987.

Hough, R. *Admirals in Collision.* London: Hamish Hamilton, 1959.

——. *The Fleet That Had To Die.* London: Hamish Hamilton, 1958.

Keegan, J. *The Face of Battle.* London: Penguin, 1978.

MacDonald, L. *1915.* London: Headline, 1993.

——. *They Called it Passchendaele.* London: Michael Joseph, 1978.

MacShane, D. *Friendly Fire Whitewash.* London: Epic, 1992.

Marshall, S. L. A. *Pork Chop Hill.* New York: Morrow, 1956.

McGuffie, T.H. *Rank and File.* London: Hutchinson, 1964.

McKay, G. *In Good Company.* Sydney: Allen and Unwin, 1987.

McManners, H. *The Scars of War,* London: Harper Collins, 1993.

Mercer, C. *Journal of the Waterloo Campaign.* London, n.p., 1877.

Lord Moran, *The Anatomy of Courage.* London: Constable 1945.

Norman, T. *The Hell They Called High Wood.* London: Patrick Stephens, 1984.

Patton, G. S. *War As I Knew It.* London: W. H. Allen, 1948.

Prior, R. and T. Wilson. *Command on the Western Front.* Oxford: Blackwell, 1992.

Pukowski, J. and J. Sykes. *Shot at Dawn.* London: Wharnecliffe, 1989.

Regan, G. B. *Someone Had Blundered.* London: Batsford, 1987.

———. *Naval Blunders.* London: Guinness, 1993.

Schwarzkopf, N. *It Doesn't Take A Hero.* New York: Bantam, 1993.

Shrader, C.R. *Amicide: The Problem of Friendly Fire in Modern War.* Fort Leavenworth: Combat Studies Institute, 1982.

Warner, D. and P. *The Tide at Sunset.* New York: Charterhouse, 1974.

Whiting, C. *Slaughter over Sicily.* London: Leo Cooper, 1992.

———. *The Battle of Hurtgen Forest.* London: Leo Cooper, 1989.

Winter, D. *Haig's Command.* London: Viking, 1991.

Wolff, L. *In Flander's Fields.* London: Penguin, 1979.

NOTES

1. Quoted in Shrader, *Friendly Fire: The Inevitable Price. Parameters,* Autumn 1992, p.29.
2. John Keegan, *The Face of Battle.* London: Penguin, 1978, pp. 100–1.
3. Froissart, *The Chronicles of England, France and Spain.* London: Dent, 1906, p.45.
4. Blaise de Monluc, *Military Memoirs* (ed. I Roy), London: Longman, 1971, p.69.
5. Christopher Duffy, *The Military Experience in the Age of Reason.* London: Routledge, 1987, p.211.
6. John Green, *A Soldier's Life,* 1806–1815. London: EP Publishing, 1818, p.17.
7. Quoted in T.H. McGuffie, *Rank and File.* London: Hutchinson, 1964, p. 44.
8. *The Recollections of Rifleman Harris.* ed. Christopher Hibbert. London: Leo Cooper, 1970, pp.26–27.
9. Quoted in T.H. McGuffie, *Rank and File.* London: Hutchinson, 1964, pp. 234–235.
10. Quoted in T.H. McGuffie, *Rank and File.* London: Hutchinson, 1964, p.254.
11. P. Griffith, *Rally Once Again.* London: Crowood Press, 1987, p. 89.

12. Quoted in P. Griffith, *Rally Once Again.* London: Crowood Press, 1987, p.89.

13. Quoted in P. Griffith, *Rally Once Again.* London: Crowood Press, 1987, p.111.

14. Quoted in *Battles and Leaders of the Civil War.* ed. R.U. Jackson and C.C. Buel, Vol. 1, New York: Castle, p.235.

15. Quoted in G. Regan, *Someone Had Blundered.* London: Batsford, 1987, p. 226.

16. Ibid., p.229.

17. Quoted in L. McDonald, *1915.* London: Headline, 1993, pp.502–503.

18. Quoted in T. Norman, *The Hell They Called High Wood.* London: Patrick Stephens, 1984, p.219.

19. Quoted in L. McDonald, *1915.* London: Headline, 1993, pp.307–308.

20. Quoted in L. McDonald, *They Called It Passchendaele.* London: Michael Joseph, 1978, p.197.

21. Quoted in L. McDonald, *They Called It Passchendaele.* London: Michael Joseph, 1978, p.198.

22. C. E. W. Bean, *Official History of Australia In The War Of 1914–18.* Vol. IV, Sydney: Angus and Robertson, p.887.

23. Quoted in J. Ellis, *The Sharp End of War.* London: David and Charles, 1980, p. 18.

24. Quoted in J. Ellis, *The Sharp End of War.* London: David and Charles, 1980, p. 267.

25. Lt. Col. C.R. Shrader, *Amicide: The Problem of Friendly Fire in Modern War.* Fort Leavenworth: Combat Studies Institute, December 1982, p.5.

26. Quoted in Lt. Col. C.R. Shrader, *Amicide: The Problem of Friendly Fire in Modern War.* Fort Leavenworth: Combat Studies Institute, December 1982, p.90.

27. Quoted in J. Ellis, *The Sharp End of War.* London: David and Charles, 1980, p.95.

28. Quoted in J. Ellis, *The Sharp End of War.* London: David and Charles, 1980, p.19.

29. Quoted in J. Ellis, *The Sharp End of War.* London: David and Charles, 1980, p.20.

30. Quoted in J. Ellis, *The Sharp End of War.* London: David and Charles, 1980, p.92.

31. G. McKay, *In Good Company.* Sydney: Allen and Unwin, 1987, p.137.

32. General N. Schwarzkopf, *It Doesn't Take A Hero.* New York: Bantam, 1993, p.232.

33. G. McKay, *In Good Company.* Sydney: Allen and Unwin, 1987, p. 114.

34. Quoted in E. Bergerud, *Red Thunder, Tropic Lightning.* Boulder: Westview, 1993, p. 130.

35. Quoted in E. Bergerud, *Red Thunder, Tropic Lightning.* Boulder: Westview, 1993, p. 141–142.

36. Quoted in E. Bergerud, *Red Thunder, Tropic Lightning.* Boulder: Westview, 1993, p. 131.

37. Quoted in E. Bergerud, *Red Thunder, Tropic Lightning.* Boulder: Westview, 1993, p. 160.

38. Quoted in E. Bergerud, *Red Thunder, Tropic Lightning.* Boulder: Westview, 1993, p. 179.

39. Quoted in H. McManners, *The Scars of War.* London: Harper Collins, 1993, p. 233.

40. Quoted in H. McManners, *The Scars of War.* London: Harper Collins, 1993, pp. 234–235.

41. C. Whiting, *Slaughter Over Sicily.* London: Leo Cooper, 1992, p.100.

42. Quoted in C. Whiting, *Slaughter Over Sicily.* London: Leo Cooper, 1992, p.129.

43. Quoted in C. Whiting, *Slaughter Over Sicily*. London: Leo Cooper, 1992, p.130.

44. Quoted in G.S. Patton, *War as I Knew It*. London: W.H. Allen, 1948.

45. Quoted in Carlo D'Este, *Decision in Normandy*. London: Collins, 1984, p.341.

46. Quoted in Max Hastings, *Overlord*. London: Michael Joseph, 1984, p.254.

47. Quoted in Carlo D'Este, *Decision in Normandy*. London: Collins, 1984, p.402.

48. Quoted in Lt. Col. C.R. Shrader, *Amicide: The Problem of Friendly Fire in Modern War*. Fort Leavenworth: Combat Studies Institute, December 1982, p. 54.

49. Quoted in G.B. Regan, *Book of Naval Blunders*. London: Guinness, 1993, p. 25.

50. Quoted in Denis and Peggy Warner, *The Tide At Sunset*. New York: Charterhouse, 1974, p.410.

51. Quoted in Denis and Peggy Warner, *The Tide At Sunset*. New York: Charterhouse, 1974, p.413.

52. Quoted in J. Baynes, *Morale: A Study of Men and Courage*. London: Cassell, 1967, p. 182.

53. Brigadier General F.P. Crozier, *The Men I Killed*. London: Michael Joseph, 1937, pp. 54–55.

54. Brigadier General F.P. Crozier, *The Men I Killed*. London: Michael Joseph, 1937, pp. 69–70.

55. Quoted in J. Putkowski and J. Sykes, *Shot at Dawn*. London: Wharncliffe Publishing, 1989, p. 304.

56. Quoted in J. Putkowski and J. Sykes, *Shot at Dawn*. London: Wharncliffe Publishing, 1989, p. 305.

57. Quoted in C. Whiting, *The Battle of Hurtgen Forest*. London: Leo Cooper, 1989, p. 170.

58. Quoted in C. Whiting, *The Battle of Hurtgen Forest.* London: Leo Cooper, 1989, p. 228.

59. Brigadier General F.P. Crozier, *A Brass Hat in No Man's Land.* London: Cape, 1930, pp. 208–209.

60. Quoted in Richard Holmes, *Firing Line.* London: Penguin, 1987, p. 330.

61. Quoted in E. Bergerud, *Red Thunder, Tropic Lightning.* Boulder: Westview, 1993, p. 134.

62. Quoted in E. Bergerud, *Red Thunder, Tropic Lightning.* Boulder: Westview, 1993, p. 303.

INDEX